Adventures of a Young Naturalist

Adventures of a Young Naturalist

The Zoo Quest Expeditions

DAVID ATTENBOROUGH

TWO
ROADS

www.tworoadsbooks.com

First published in Great Britain in 1980 by Lutterworth Press as
The Zoo Quest Expeditions: Travels in Guyana, Indonesia and Paraguay

This edition first published in Great Britain in 2017 by
Two Roads
An imprint of John Murray Press
An Hachette UK company

11

A CIP catalogue record for this title
is available from the British Library

Hardback ISBN 978-1-47366-440-1
Trade Paperback ISBN 978-1-47366-595-8
Ebook ISBN 978-1-47366-441-8

Typeset in Bembo by Palimpsest Book Production Ltd, Falkirk, Stirlingshire

Printed and bound by Clays Ltd, St Ives plc

Hodder & Stoughton policy is to use papers that are natural, renewable
and recyclable products and made from wood grown in sustainable forests.
The logging and manufacturing processes are expected to conform to the
environmental regulations of the country of origin.

Hodder & Stoughton Ltd
Carmelite House
50 Victoria Embankment
London EC4Y 0DZ

www.hodder.co.uk

Contents

Introduction   vii

BOOK ONE: Zoo Quest to Guyana

1. To Guyana   3
2. Tiny McTurk and Cannibal Fish   15
3. The Painted Cliff   29
4. Sloths and Snakes   43
5. Spirits in the Night   59
6. Shanties on the Mazaruni   80
7. Vampires and Gertie   93
8. Mr King and the Mermaid   109
9. Return   125

BOOK TWO: Zoo Quest for a Dragon

10. To Indonesia   131
11. The Faithful Jeep   142
12. Bali   157
13. The Animals of Bali   165
14. Volcanoes and Pickpockets   175
15. Arrival in Borneo   188
16. Charlie, the Orangutan   203
17. A Perilous Journey   217
18. The Island of Komodo   237
19. The Dragons   251
20. Postscript   262

BOOK THREE: Zoo Quest in Paraguay

21. To Paraguay   267
22. The Decline of a Luxury Cruise   272
23. Butterflies and Birds   286
24. Nests on the Camp   311
25. Beasts in the Bathroom   325
26. Chasing a Giant   337
27. Ranch in the Chaco   349
28. Chaco Journey   363
29. A Second Search   377
30. Moving a Menagerie   388

Introduction

These days zoos don't send out animal collectors on quests to bring 'em back alive. And quite right too. The natural world is under more than enough pressure as it is, without being robbed of its most beautiful, charismatic and rarest inhabitants. Now most of a zoo's crowd-attracting species – lions, tigers, giraffes and rhinoceros, even lemurs and gorillas – have been born in zoos and kept track of in stud books, so that individuals can be exchanged internationally without incurring problems of in-breeding. They can then play a valuable part in familiarising visitors with the splendours of the natural world and in explaining the importance and complexities of conservation.

But it was not always so. London Zoo was founded in 1828 by men of science who were, at that time, still concerned with the important but almost impossible task of compiling a catalogue of all the species of animals alive today. Some were sent to it from distant parts of the world as dead specimens. Others arrived alive and were put on display in the Society's gardens in Regent's Park. But both kinds ended up as well-studied anatomical specimens and carefully preserved. Needless to say, special attention was paid to finding species that no other zoo had ever possessed, and that ambition, to some extent, still lingered on even in the 1950s when I visited one of the Zoo's curators with an idea for a new kind of television programme.

Television then was also very different from what it is today. There was only one network, produced by the BBC, which could only be seen in London and Birmingham. All its programmes came from two small studios in Alexandra Palace, in north London. They were the same studios and indeed the same cameras that in 1936 had provided the first regular television service in the world. Transmissions were suspended in 1939 on the outbreak of the Second World War, but then resumed as soon as peace was declared in 1945. So when, in 1952, I got a job as a trainee producer, British television had only ten years of practical production experience.

The programmes were almost entirely live. Electronic recording was still decades away, so the only way we producers had of supplementing the pictures from the studio was with film. That cost money and we were seldom given enough to enable us to do so. This was not regarded as much of a limitation. On the contrary, both viewers and producers thought that the 'immediacy' was the medium's main attraction. The events appearing on the screen were actually happening as viewers watched them. If an actor forgot his lines, then the prompt was audible. If a politician lost his temper, then all saw him do so and there was no chance for him to have second thoughts and insist that his incautious words were edited out.

Animal programmes were already established in the schedules when I first started. They were presented by George Cansdale, the Superintendent of the London Zoo. Week after week, he transported some of the more reasonably-sized and amenable of his charges from Regent's Park to Alexandra Palace and put them on a table covered with a doormat, where they sat blinking in the intense lights of the studio, while Mr Cansdale demonstrated their anatomy, their bravery and their party tricks. He was an expert naturalist, marvellously adept at handling animals and persuading them to do what he wanted. Even so things did not always go as he might have wished. That was part of his popularity. They regularly relieved themselves on the doormat or, with luck, over his trousers. Occasionally they escaped and had to be fielded by one of the uniformed zoo keepers lurking in the wings ready for such an eventuality. Once a small African squirrel leapt from the demonstration table on to the microphone, hanging on a boom directly above. From there it scampered across the studio and found refuge in the ventilation system. It lived there for days, making occasional appearances in the dramas, variety shows and epilogues that continued to come from its studio. On a few memorable occasions an animal even managed to give Mr Cansdale a nip. Such a moment was not to be missed and when he produced a particularly dangerous creature, like a snake, the nation held its breath.

Then, in 1953, a new kind of animal programme appeared. A Belgian explorer and film-maker called Armand Denis, together with his glamorous British-born wife Michaela, came to London from Kenya to publicise a feature-length documentary they had made for the cinema called *Below the Sahara*. They had put some of the footage they had not used in the film into a half-hour programme for television. It showed elephants, lions, giraffe and much of the rest of the famous and spectacular big game of the east African plains. It was a huge success. For many viewers it was the first time they had seen moving pictures of such creatures. Although the images did not have the live, titillating unpredictability of those produced by Mr Cansdale, people could see how marvellous and majestic the animals were in their proper setting.

Viewers responded with such enthusiasm to this new kind of animal programme that the television planners immediately asked the Denises for more. Why not a whole series that could run week after week? The Denises, who had been filming in Africa for years and had a vast library of animal footage, saw the possibilities and needed little persuasion. So the first *On Safari* series began.

For me, a twenty-six-year-old novice television producer with two years' broadcasting experience and an unused zoology degree, anxious to make animal programmes myself, it seemed that each of these formats had its own particular attractions – and its own limitations. Mr Cansdale's had the undeniable thrill of watching living unpredictable animals, but since the animals were always in an alien studio environment, they looked, more often than not, bizarre oddities. The Denises' animals, on the other hand, appeared in the natural surroundings to which they were perfectly adapted, but they lacked the spice of live unpredictability. Surely, I argued to myself, it should be possible to combine the two styles in one programme and get the benefits of both. I had already, as a jobbing producer, directed music recitals, archaeological quizzes, political discussions and ballet performances. Most recently, I had devised a series of three programmes about the meaning and purpose of animal shapes and patterns. They had been narrated by one of the great scientific figures of the times, Sir Julian Huxley, and, to illustrate his words, I had borrowed some of the animals from Mr Cansdale's London Zoo. And in doing that I had met the Zoo's Curator of Reptiles, Jack Lester.

Jack, from his earliest days, had had a passion for animals, but lacking any formal training, he had first taken a job in a bank. However, he soon persuaded his employers to send him to a branch in West Africa, and there he was able to indulge his enthusiasm for collecting and keeping reptiles. When the war came he joined the Royal Air Force, but after it was over he got a job in a private zoo in the west of England. From there he had come to Regent's Park to care for the Zoo's large collection of reptiles. His office was a small room in the Reptile House, heated, like all the showcases, to a suffocating tropical temperature and filled with all kinds of cages containing his particular favourite creatures that were not needed for public exhibition – dwarf bush babies, giant spiders, chameleons and burrowing snakes. He had been a great help in selecting animals for the Huxley series, and I went there to discuss what more programmes we might do together. I thought I had an idea that might interest him, because it would get him back to his beloved West Africa – and I with him.

My plan was simple. The BBC and the London Zoo should mount a joint animal-collecting expedition on which we should both go. I would direct film sequences showing Jack searching for and finally capturing a creature of particular interest. The sequence would end with a close-up of the animal in his hands. The picture would then dissolve into a similar shot of the same creature, but this time live in the studio. Jack would then demonstrate, in the Cansdale manner, the particularly interesting aspects of its anatomy and behaviour. If there were a few unavoidable incidents, such as an escape or a bite, then so much the better. Viewers would then be returned, by way of film, to Africa and another search and capture.

Jack agreed that the idea was a good one. The only problem was that the Zoo at the time had no intention of sending out a collecting expedition. Nor had the BBC any intention of embarking on the highly specialised and certainly expensive business of making natural history films. That minor difficulty, however, might be overcome with a single properly stage-managed lunch at which bosses from both the Zoo and the BBC would meet, under the impression that the other already had such a scheme in mind.

The lunch duly took place in the Zoo's restaurant. Jack and I were there to prompt and steer our seniors. Both bosses left after coffee, each convinced that his organisation had a lot to gain from joining in the other's plans, and to our incredulous delight the very next day we were both told separately to go ahead.

We agreed on the jungle without any difficulty. Jack's bank had been in Sierra Leone. He knew the country and he knew the fauna. He still had a lot of friends there who could give us help. I was convinced, however, that if the television programmes were to be a success, the expedition should have one particular objective – a rare creature that, since the Zoo was excited by such things, had never been seen in any zoo anywhere else in the world; an animal so romantic, rare and exciting that the quest for it would keep viewers watching programme after programme until in the last the animal was finally found. We could call the series 'Quest for . . .' something . . . But what?

It was a difficult bill to fill. The only animal that Jack could think of in Sierra Leone that might remotely qualify was a bird called *Picathartes gymnocephalus*. It seemed to me that rousing the British public into a frenzy of excited anticipation to see a creature with such a name might be difficult. Had it not got another, more romantic one? 'Yes, indeed,' Jack said helpfully, 'its English name is Bare-headed Rock Fowl.' Even that, I thought, would scarcely do the job. But Jack could think of no other. So *Picathartes* became our ultimate target and I decided to call the series simply *Zoo Quest*.

There was a further issue to be settled. The film used by television at that time was 35 mm wide, the same as that used by the feature-film industry. A single roll was about the size of a flattened football, and the camera that used it was as big as a small suitcase. In normal circumstances it needed to be mounted on a tripod with two people to handle it. Armand and Michaela Denis had used a much smaller piece of equipment that used 16 mm film, and I wanted to do the same.

The head of the television film department was outraged: 16 mm, he said, was for amateurs. Professionals despised it. Its pictures were unacceptably fuzzy. He would rather resign than agree to such lowering of standards. A meeting was called by the head of television programmes. I put my case. I could not get the shots I needed, I explained (with the confidence of having never done anything of the kind), unless I used the much smaller, more easily handleable equipment.

Eventually the argument went my way. But in accepting the decision, the head of films had a proviso. Television at that time was black and white. The 16 mm film we should use, however, should not be black-and-white negative from which positive prints could be taken, but colour negative. This was not as sensitive as some black-and-white stocks, but black-and-white prints taken from it had much better resolution. I accepted and agreed that we would only use black-and-white negative in very exceptional circumstances where the light was extremely dim.

None of the BBC's staff cameramen, however, would agree to use 16 mm equipment. I would have to find a cameraman for myself. I made a few enquiries and discovered that a man of my own age had just returned from the Himalayas where he had been the assistant cameraman filming an expedition looking (unsuccessfully) for the Abominable Snowman. His name was Charles Lagus. We arranged to meet in a pub close to the studios where television people habitually met. We drank a little beer. We laughed at the same jokes. He thought the trip sounded fun and after a second drink, he agreed to come. Jack too enlisted a recruit, Alf Woods, a wily and sagacious head keeper who was then in charge of the Zoo's Bird House. He would have the job of caring for the animals as they were caught. So, in September 1954, the four of us departed for Sierra Leone.

After a few days in Freetown, the country's capital, we set off for the rainforest. Neither Charles nor I had ever been in such a place before. It was extraordinarily dim. Charles gloomily took out his light meter. 'The only way we can get enough light to shoot colour neg. here,' he said bitterly, 'is to cut down a couple of trees.' It was a serious blow. If we were going to work in the forest, we would have to use the black-and-white stock and we had a very little of it.

Could we perhaps persuade Jack, once he had caught something in the forest, to release it in a suitably sunny clearing and

Alf Woods (right) and Jack Lester feeding the Picathartes *chick*

then catch it a second time? Jack gallantly agreed to do so. And Charles and I, instead of trying to film troops of monkeys gambolling through the branches or waiting in hides for a shy forest antelope to appear out of the gloom, would limit ourselves to those small creatures which we too could take out into the light – chameleons, scorpions, mantises and millipedes.

Picathartes would remain our main target. Jack had brought with him a small watercolour drawing of it that an artist had made from a museum specimen, and wherever we went he showed it to people asking if they had ever seen anything like it. People looked at it in bafflement, but eventually we found a villager who recognised it. The bird, he said, built mud nests, rather like those of swallows only very much larger, which it attached to the flanks of great stone boulders buried in the forest. They could hardly be moved into the light. Nor did we try to improve our chances by felling a few nearby trees. Instead we used our precious hypersensitive black-and-white film and at last succeeded in getting the first pictures ever taken of a living *Picathartes gymnocephalus*.

The first programme reached television screens in December 1954. Jack showed the animals in the studio and I directed the cameras and cued the film sequences from the control gallery. But there was a great sadness. The day after the transmission, Jack collapsed and was taken to hospital. The series was, of course, live. So the following week, someone else had to take his place. The head of television instructed me to do so. 'You are on the staff,' he said, 'so there will be no extra fee.' The following week I did my best to take Jack's place, handling the animals, while one of my director friends sat in the control gallery directing the cameras.

The Africa we showed was very different from the one presented by the Denises. Potter wasps building their astonishing cup-shaped nests and columns of army ants attacking a scorpion were much smaller than the big game of east Africa, but Charles, with his skilful photography, had made them look extremely dramatic and the series attracted a remarkably big audience. My bosses were delighted.

A month or so after the series finished, Jack had recovered sufficiently to be discharged from hospital. He and I got together again and decided that we should propose another series while our respective bosses still remembered that the first had been a success.

So we did and – rather to our surprise – in March 1955, a mere eight weeks after the last West African programme had been transmitted, we set off again, this time for South America and what was then called British Guiana.

But soon after we had arrived there Jack's illness returned, and he had to fly back to hospital in London. So, once again, I had to take on his role, this time as an animal collector, and another head keeper from the Zoo came out to look after the animals as the collection grew.

Jack had still not properly recovered by the time we got back and I once again presented the series. It, too, was a success so we proposed a third trip. This time we decided on Indonesia where our main target would be the Komodo dragon, the largest lizard in the world, that then had never been seen on television. Jack would plainly not be fit enough to travel, but he urged us to go without him. So we did. He died while we were away at the tragically early age of forty-seven.

After the Guiana trip I wrote an account of our experiences and I did so after each of the others that followed annually for the next few years. This book contains the first three, slightly abbreviated and updated from the originals.

The world has changed a great deal since they were written. British Guiana has become independent and taken the name of Guyana. The Rupununi savannahs where we went to look for giant anteaters that then seemed so wild and remote to us are now served by a regular air service and have easy communications with the coast. In Indonesia, the great Javanese monument of Borobudur, which then was romantically falling into ruin, has now been completely dismantled and reconstructed; Bali, which we could only reach by sea and where we only saw one other European face, today has an airfield where giant jets daily bring thousands of holidaymakers on their journeys between Australia and Europe; and Komodo which we reached with such difficulty in 1956 is today on the tourist route and parties of visitors are taken daily to view the dragons. And television itself since those days at last transmits its images in colour.

In 2016, however, an archivist, sorting through the contents of one of the BBC's film vaults, discovered a few rusting cans labelled 'Zoo Quest – Colour'. Puzzled, she opened them and discovered rolls of the original colour negative which, until then no one, including myself, had ever viewed in colour. So they were at last printed in colour. Those that viewed them decided that, after their sixty-year hiccup, they had such vividness that they should be shown on air. I hope the pages that follow may achieve a similar sort of thing.

David Attenborough, May 2017

BOOK ONE

Zoo Quest to Guyana

I

To Guyana

———

South America is the home of some of the strangest, some of the loveliest and some of the most horrifying animals in the world. There can be few creatures more improbable than the sloth which spends its life in a permanent state of mute slow motion, hanging upside down in the tall forest trees; few more bizarre than the giant anteater of the savannahs with its absurdly disproportionate anatomy, its tail enlarged into a shaggy banner and its jaws elongated into a curved and toothless tube. On the other hand, beautiful birds are so common as to become almost unremarkable: gaudy macaws flap through the forest, their splendid plumage contrasting incongruously with their harsh maniac cries; and hummingbirds, like tiny jewels, flit from flower to flower sipping nectar, their iridescent feathers flashing the colours of the rainbow as they fly.

Many of the South American animals inspire the fascination which comes from revulsion. Shoals of cannibal fish infest the rivers waiting to rip the flesh from any animal which tumbles among them, and vampire bats, a legend in Europe but a grim reality in South America, fly out at night from their roosts in the forest to suck blood from cows and men.

I had no doubt that, since we had visited Africa for our first Zoo Quest expedition, South America was the obvious choice for our second. But which area in such a vast and varied continent should we visit? Eventually we selected Guyana (then known as British Guiana), the only Commonwealth country in the whole of the South American continent. Jack Lester, Charles

3

Lagus and I, who had been together in Africa were to go again, and we were to be joined by Tim Vinall, one of the overseers in the London Zoo. His current responsibility was the care of the hoofed animals, but during his long career in the Zoo he had looked after many other types of creature. His was to be the back-breaking and thankless task of remaining at our base at the coast and looking after the animals as we caught them and brought them to him.

So in March 1955, we landed in Georgetown, the capital. After three days of obtaining permits, clearing our cameras and recording apparatus through customs, and buying pots and pans, food and hammocks, we were itching to begin our collecting in the interior. We had already decided on an approximate plan of action. From the map we had seen that most of Guyana is covered by tropical rain forest which extends northwards to the Orinoco and southwards to the Amazon Basin. In the south-west, however, the forest dwindles and gives way to rolling grass-covered savannahs, and lining the coast is a strip of culti-vated land where rice fields and sugar plantations alternate with swamps and creeks. If we were to assemble a representative collection of the animals of Guyana, we should have to visit each of these areas, for each harbours creatures which are not to be found elsewhere. We had little idea, however, where we should go in each of the districts and in what order to visit them, until on our third evening we were invited to dinner with three people who could give us expert advice: Bill Seggar, a District Officer in charge of a remote territory in the forests near the far western frontier, Tiny McTurk, a rancher from the Rupununi savannahs, and Cennydd Jones, whose work as doctor to the Amerindians took him to every corner of the colony. We sat up until early in the morning looking at photographs and films, poring over maps and excitedly scribbling notes. When we finally broke up, we had decided upon a detailed campaign, visiting first the savannahs, next the forest, and finally the coastal swamps.

The following morning, we walked into the Airways office to inquire about transport.

'The Rupununi for four, sir?' said the clerk. 'Certainly. A plane is leaving tomorrow.'

Charles Lagus with a matamata turtle

It was with a sense of great excitement that Jack, Tim, Charles and I clambered into the plane which was to take us there. Nevertheless, we did not expect to find our hearts in our mouths as soon as we did. Our pilot, Colonel Williams, had pioneered bush flying in Guyana and it was largely through his daring and imagination that many of the remoter parts of the country had become accessible at all. As we took off, however, we discovered that the Colonel's flying technique was very different from that of the pilot who had brought us from London to Georgetown. Our Dakota thundered down the airstrip; the palm trees at the end loomed nearer and nearer, until I thought that something was wrong with the machine and that we were unable to leave the ground. At the very last moment we surged into the air in a steep climb, missing the tops of the palm trees by feet. We all exchanged ashen looks, and after shouting our doubts and worries to one another, I went forward to ask Colonel Williams what had happened.

'In bush flying,' he yelled, out of the corner of his mouth, tapping his cigarette into the tin ashtray tacked on to the control panel, 'in bush flying, I reckon the most dangerous time is at take-off. If one engine fails then, when you are needing it most, you land with a crash in the forest and there's no one there to help you. I always reckon to get up so much speed on the ground that my momentum is enough to take me up on no engines at all. Why, boy, are y'all scared?'

I hastily reassured Colonel Williams that none of us had been in the least bit worried; we were merely interested in the technique of handling aircraft. Colonel Williams grunted, changed the short-focus spectacles that he had worn for the take-off for a long-focus pair, and we settled down for the flight.

Beneath us stretched the forest, a green, velvet blanket spreading as far as we could see in all directions. Slowly it began to rise towards us as we approached a great escarpment. Colonel Williams flew on without altering height until the forest came so close to us that we could see parrots flying above the trees. Then as the escarpment fell away, the forest began to change character. Small islands of grassland appeared and soon we were flying over wide open plains veined with silver creeks and freckled with tiny, white termite hills. We lost height, circled over a small cluster of white buildings and shaped up for a landing on the airstrip – a euphemism for a stretch of the savannah which seemed to differ from its surroundings only in that it was clear of termite hills. The Colonel brought the plane down gracefully, and bumpily taxied towards a little knot of people awaiting the plane's arrival. We clambered over the piles of freight lying lashed on the floor of the Dakota and jumped out, blinking in the brilliant sun.

A cheerful, bronzed man in shirt sleeves and sombrero detached himself from the onlookers and came over to meet us. It was Teddy Melville, who was to be our host. He came from a famous family. His father was one of the first Europeans to settle on the Rupununi and begin ranching the cattle that were now thinly spread throughout the district. He arrived at the turn of the century and married two Wapishana girls who each presented him with five children. These ten men and women now occupied nearly all the important positions in the district; they were ranchers, store keepers, government rangers and hunters. We soon discovered that, no matter where we went in the northern savannahs, if the man we met was not a Melville, then as like as not he was married to one.

Lethem, where we had landed, consisted of a few white concrete buildings, untidily scattered round two sides of the airstrip. The largest of them, and the only one to have an upper storey, was Teddy's guest-house – a plain rectangular building with a veranda and gaping glassless windows, which was graced by the title of Lethem Hotel. Half a mile away to the right, on the crest of a low rise, stood the District Commissioner's house, the post office, a store and a small hospital. A dusty red-earth road ran from them to the hotel and continued past a group of ramshackle outhouses into a parched wilderness of termite hills and stunted bushes. Twenty miles beyond, jutting abruptly from the plains, rose a line of jagged mountains, reduced by the heat-haze to a smoky-blue silhouette against the dazzling sky.

Everyone for miles around had come to Lethem to meet the plane, for it brought with it long-awaited stores and the regular weekly mail. Plane days therefore were always great social occasions, and the hotel was crowded with ranchers and their wives who had driven in from outlying districts and who remained after the plane had left to exchange news and gossip.

After the evening meal was over, the bare deal tables were cleared from the dining-room and long wooden benches set in their place. Harold, Teddy's son, began setting up a film projector and a screen. Gradually the bar emptied and the benches were filled. Wapishana cowboys, known as vaqueros, bronzed with straight blue-black hair and bare feet, trooped in and paid at the door. The air was filled with rank tobacco smoke and expectant chatter as the lights were put out.

The entertainment began with some sensibly undated newsreels. These were followed by a Hollywood cowboy film about pioneering the wild west, during which virtuous white Americans convincingly slaughtered great numbers of villainous Red Indians. Hardly tactful one would have thought, but the Wapishana sat watching their North American cousins being exterminated without any emotion on their impassive faces. The story was a little difficult to follow for not only had lengthy sequences been excised during the copy's long life, but it seemed doubtful whether the reels were projected in their correct order, for a tragic and beautiful American girl who was savagely murdered by the Indians in the third reel, reappeared in the fifth to make love to the hero. But the Wapishana were an accommodating audience, and a pedantic detail of this kind did not spoil their obvious enjoyment of the big fight scenes, which provoked rounds of enthusiastic applause. I suggested to Harold Melville that the film was perhaps an odd choice, but he assured me that cowboy films were by far the most popular type of any they showed. Certainly one could believe that Hollywood bedroom comedies would seem even greater nonsense to the Wapishana out here.

After the show, we went upstairs to our room. In it were two beds equipped with mosquito nets. Two of us obviously had to sleep in hammocks, and Charles and I claimed the privilege. It was an opportunity which both of us had been thirsting to seize ever since we had bought our hammocks in Georgetown. With a highly professional air we slung them from hooks fastened in the walls. The results however, as we realized after a few weeks of experience, were hopelessly amateur. We had hitched them far too high and had tied them with enormously elaborate knots that were going to take a considerable time to loosen in the morning. Jack and Tim stolidly climbed into their beds.

The next morning there was little doubt as to which pair of us had spent the more comfortable night. Charles and I both swore that we had slept like logs and that sleeping in hammocks was second nature to us. But it was hardly true, for neither of us had then learnt the simple technique of lying diagonally across the stretcherless South American hammock. I had spent most of the night trying to lie along the length of it, with the result that my feet were higher than my head and my body was slumped in a great curve. I had been unable to turn without breaking my back, and I got up that morning feeling that I should be afflicted with a permanent curvature of the spine.

After breakfast, Teddy Melville came in with the news that a large party of Wapishana had started fishing in a nearby lake by the traditional method of poisoning its waters. There was a chance that in the process they would come across other animals which might be of interest to us, and Teddy suggested we should go over to have a look. We got into his truck and set off across the savannahs. There was little to prevent us from driving wherever we wished. Here and there were tortuously weaving creeks, but they were easily avoided; we could see them from a considerable distance away, their banks being fringed by bushes and palm trees. Otherwise the only obstacles in our way were clumps of stunted sandpaper bushes and termite hills – tall, crazily spired towers, sometimes standing singly and sometimes concentrated in groups so dense that at times it seemed we were driving through a giant graveyard. A few well beaten tracks across the savannahs linked one ranch to the next, but the lake we were to visit was isolated and before long Teddy branched off the main trail and began threading bumpily between the bushes and the termite hills, following no track but simply relying on his sense of direction. Soon we saw a belt of trees on the horizon marking the site of the lake we were to visit.

When we arrived, we found that a long arm of the lake had

been dammed with a barricade of stakes. Into it the Wapishana had crushed special lianas which they had gathered many miles away in the Kanuku mountains. All around were fishermen with bows and arrows at the ready, waiting for the fish to become stupefied by the poisonous sap of the lianas and float to the surface. The Wapishana clung to branches of trees overhanging the lake's margin; they perched on specially built platforms in the middle of the water; some stood on small improvised rafts and others patrolled up and down in dugout canoes. In a clearing on the bank the women had lit fires and slung hammocks and now sat waiting to clean and cure the fish as soon as the men brought them in; but nothing so far had been caught and the women were getting impatient. Their menfolk had been foolish, they said scornfully: too big a section of the lake had been dammed and too few lianas had been gathered in the forest, so that the poison was too weak to affect the fish. Three days of hard work in damming and platform building had been wasted. Teddy talked to them in Wapishana and gathered all this information as well as the news that one of the women had seen a hole in the bank on the other side of the lake, which she said was occupied by a large animal. What kind of animal it was, she was not sure; it might be either an anaconda or a caiman.

The caiman belongs to the same group of reptiles as the crocodile and alligator, and to the layman all three animals look very much alike. To Jack, however, they were very different, and though all three are found in the Americas, they each have distinctive habitats. Here on the Rupununi, Jack said, we could expect to find the black caiman, the largest species in its own group, which is reputed to grow up to twenty feet long. Jack admitted that he would rather like a 'nice big caiman' and, come to that, he would also be quite glad to catch a sizeable anaconda. As the animal in the hole might turn out to be one or the other, he felt we really should try to catch it. We all climbed into dugout canoes and paddled across the lake with one of the women to guide us.

Shooting fish

On investigation, we found that there were two holes – a small one and a large one, and that they were connected with each other, for a stick pushed down the smaller one provoked splashes from the other. We barricaded the smaller hole with stakes. To prevent the unknown creature from escaping through the larger one and, at the same time, to allow it enough space to emerge and be caught, we cut saplings from the bank and drove them deep into the mud of the lake bottom in a semi-circular palisade around the entrance. We had not yet seen our quarry and no amount of prodding through the smaller hole would drive it out, so we decided to enlarge the big hole by cutting through the turfy bank. Slowly we hacked away the roof of the tunnel, and as we did so the bank shook with a subterranean bellow that could hardly have been produced by a snake.

Cautiously peering through the stakes of the palisade into

a gloomy tunnel, I just distinguished, half submerged in the muddy water, a large yellow canine tooth. We had cornered a caiman, and judging from the size of the tooth, a very large one.

Digging out the caiman

A caiman has two offensive weapons. First and obviously, its enormous jaws; and second, its immensely powerful tail. With either it can inflict very serious injuries, but fortunately the one we were tackling was so placed in its hole that we only had to pay attention to one end at a time. Having had that momentary glimpse of its teeth, I knew which end was uppermost in my mind. Jack was paddling about in the muddy water inside the stakes trying to work out how the caiman was lying and how best to tackle the job of catching it. It seemed to me that if the beast elected to come out in a hurry, Jack would have to jump very quickly to avoid losing a leg. For my part, I felt I was quite near enough to danger wading thigh-deep farther out in the lake, manoeuvring Charles in a canoe at a sufficient distance to get good film shots of the proceedings. In the event of the caiman making a lunge at Jack, I was quite sure that it would come with such a rush that it would knock our flimsy palisade flat, and whereas Jack could leap for the bank, I should have to wade several yards before I reached safety. I was in no doubt that the caiman, in such a depth of water, would be able to move faster than me. For some reason or other – perhaps my nervousness showed itself more than I imagined – I seemed unable to keep the canoe steady enough to make it practicable for Charles to work, and after I had given it a particularly violent lurch, which nearly threw him and his camera into the water, he decided that his apparatus would stand less chance of getting wet if he joined me wading in the lake.

Meanwhile, Teddy had borrowed a rawhide lasso from one of the Wapishana, and he and Jack, kneeling on the bank, were dangling it in front of the caiman's nose in the hope that it might lunge forward towards Charles and me and, in doing so, thrust its head through the noose. It roared and thrashed the sides of its tunnel so violently that the whole bank quivered, but very sensibly it refused to come out any further. Jack cut more of the bank away.

By now there were some twenty locals watching the proceedings and offering suggestions. To them it seemed incomprehensible that we should wish to catch the creature alive and unharmed. They were in favour of despatching it there and then with their knives.

At last, with the aid of two forked sticks to hold the noose wide open, Jack and Teddy coaxed the lasso round the caiman's black snout. This plainly infuriated the beast and with a twist and a roar it shook the noose off. Three times the rope was on and three times it was shaken off. It went round a fourth time. Slowly, with the sticks, Jack eased it up towards the caiman's head. Then suddenly, before the reptile realized what was happening, he drew the noose tight and the dangerous jaws were secured.

Now we had to guard against a blow from its huge tail. The situation began to look more alarming from where Charles and I were standing, for, having tied another noose round the caiman's jaws for safety, Teddy told the Wapishana to uproot the palisade. There was nothing now but open water between Charles and me and the caiman which lay with its long head projecting out of the hole, glaring at us malevolently with yellow unblinking eyes. Jack, however, jumped down from the bank into the water immediately in front of the hole, taking with him a long pole he had cut from a sapling. Bending down, he pushed the pole into the tunnel so that it lay along the reptile's scaly back, and reaching inside he secured it by tying a half-hitch round the pole and under the animal's clammy armpits. Teddy joined him and, inch by inch, they drew the caiman out of its hole, tying half-hitches around its body and on to the sapling as it emerged. The back legs, the base of the tail, and finally the tail itself were securely tied and the animal lay safely trussed at our feet, the muddy water lapping round its jaws. It was just ten feet long.

It now had to be ferried across the lake to the trucks. We hitched the front end of the pole to the stern of a dugout canoe, and towing the caiman behind us we paddled back to the women's encampment.

Jack supervised the Wapishana as they helped us to load the caiman on to the truck and then he methodically inspected its bonds one by one to see that none was chafing. The women, having no fish to cure, gathered round the truck, examining our capture and trying to decide why on earth anybody should value such a dangerous pest.

We drove off back across the savannahs. Charles and I sat on each side of the caiman with our feet within six inches of its jaws, trusting that the rawhide lassoes were as strong as they were reputed to be. We were both jubilant at having caught such an impressive creature so early. Jack was less demonstrative.

'Not bad,' he said, 'for a start.'

2

Tiny McTurk and Cannibal Fish

———

After a week on the savannahs we found, rather to our surprise, that we had assembled quite a large menagerie. We had captured a giant anteater, the vaqueros had brought us many kinds of animals and Teddy Melville had contributed by giving us several of the pets that roamed about his house – Robert, a raucous macaw, two trumpeter birds which had been living semi-domesticated lives among the chickens, and Chiquita, his capuchin monkey, who, though very tame, had the trying habit of slyly stealing things from our pockets when we were innocently playing with her.

With our collection of animals well established in Tim's care, we decided to extend our search beyond the immediate neighbourhood of Lethem, and to visit Karanambo, sixty miles away to the north. Karanambo was the home of Tiny McTurk, the rancher who had invited us to stay with him when we had met him on our third day in Georgetown. We said goodbye to Tim, climbed into a borrowed jeep, and set off.

After three hours' driving through the scrubby featureless savannahs, we saw on the horizon a belt of trees lying across the line of the trail we were following. There was no sign of a gap or clearing to suggest there was a way through and it looked as if the track must dwindle and peter out. We were sure that we had lost our way, but then we saw that the path plunged straight into the trees, down a narrow gloomy tunnel just wide enough to admit our jeep. The tree trunks on either side were interwoven with small bushes and lianas, and branches met overhead to form an almost solid ceiling.

Then unexpectedly, sunshine flooded down on us. The belt
of bush ended as suddenly as it had begun and in front of us
was Karanambo: a group of mud brick and thatched houses,
sprinkled around a wide, gravelled clearing and interspersed with
groves of mangoes, cashews, guavas and lime trees.

Tiny and Connie McTurk had heard the jeep and had come
out to greet us. Tiny was tall and fair and dressed in an oily
khaki drill shirt and trousers, for we had interrupted him in his
workshop where he was fashioning new iron arrowheads. Connie,
shorter, slim and neat in blue jeans and a blouse, greeted us
warmly and showed us into the house. We then entered one of
the most curious rooms I have ever visited. It seemed to contain
a world of its own, the old and primitive, and the new and
mechanical – a microcosm of life in this part of the world.

Room, perhaps, is not an entirely accurate word, for on two
adjoining sides it was open to the sky, the bounding walls being
only two feet high. Straddling the top of one of them was a
leather saddle, and just outside a long wooden rail carried four
outboard engines. Behind the wooden walls on the other two
sides of the room lay the bedrooms. A table against one of these
walls was covered with radio apparatus, with which Tiny maintained contact with Georgetown and the coast, and by the side
of it stood a large set of shelves crammed with books. On the
other wall hung a large clock and a barbaric assortment of guns,
crossbows, longbows, arrows, blowpipes, fishing lines and a
Wapishana feather headdress. In the corner, we noticed a stack
of paddles and an Amerindian earthenware jar full of cool water.
In the place of chairs there were three large gaily-coloured
Brazilian hammocks slung across the corners of the room, and
in the centre, its feet embedded deep in the hard-packed mud
floor, stood a giant table about three yards long. Above us, on
one of the beams, hung a line of orange-coloured maize heads,
and, here and there, stretching across the beams, a few planks
provided a spasmodic semblance of a ceiling. We looked around
admiringly.

'Not a nail in the place,' said Tiny proudly.

'When did you build it?' we asked.

'Well, after the Great War I messed about in the interior, washing for diamonds in the north-west, hunting, digging for gold and that sort of thing, and then I thought it was time I settled down. I had already made one or two trips up the Rupununi River. In those days, we did it by boats up the rapids, and it took us sometimes a fortnight and sometimes a month according to the state of the river. I thought it was a nice sort of country – not too many people, you know – and I decided to make it my home. I came up the river looking for a place that was on high ground – so that I should be above the kaboura flies and wouldn't have difficulties with drainage – and which was also near enough to the river to enable me to bring all my stores and things up from the coast by boat. Of course, this house is really only a temporary one. I put it up in rather a hurry while I was laying out the plans and getting up all the materials to build a really fancy residence. I have still got all the plans in my mind and all the materials in the outhouse and I could start building it tomorrow, but somehow,' he added, avoiding Connie's eye, 'I don't ever seem to get started on it.'

Connie laughed. 'He's been saying that for twenty-five years,' she said, 'but y'all will be hungry, so let's sit down and eat.' She moved over to the table and motioned to us to sit down. Around the table there were five up-ended orange boxes.

'I apologize for those terrible old things,' said Tiny. 'They're not nearly as good as the orange boxes we used to get before the war. You see, we once had chairs, but this floor is rather uneven and the chairs were always breaking their legs. Boxes haven't got any legs to break, so they last much longer, and really they are just as comfortable.'

Meals with the McTurks were rather complicated. Connie had the reputation of being one of the finest cooks in Guyana and certainly the meal she put in front of us was magnificent. It started with steaks of lucanani, a delicate-tasting fish which

17

Recording Tiny and Connie McTurk

Tiny regularly caught below the house in the Rupununi River. Roast duck followed – Tiny had shot them the previous day – and the meal ended with fruit from the trees outside. But competing for the food were two birds; a small parakeet and a black and yellow hangnest. They flew on to our shoulders begging for titbits, and as we were slightly unsure as to the correct way of behaving under these circumstances, we were a little slow in selecting morsels from our plates for the birds. The parakeet therefore decided to dispense with ceremony, perched on the rim of Jack's plate and helped herself. The hangnest adopted a different procedure and gave Charles a severe peck on the cheek with her needle-sharp bill to remind him of his responsibilities.

Connie, however, soon put a stop to this, chased the birds away and provided a specially cut-up meal for them in a saucer at the far end of the table. 'That's what comes of breaking rules and feeding pets at the table. Your guests are pestered,' she said.

As dusk fell towards the end of the meal, a colony of bats began to wake in the store-room and, leisurely and silently, flit across the living-room and out into the evening to begin hawking for flies. There was a scrabbling noise in the corner. 'Really, Tiny,' said Connie severely, 'we must do something about those rats.'

'Well, I did!' replied Tiny, a little hurt. He turned to us. 'We had a boa-constrictor living in the passage which used to keep the place absolutely free from rats and then just because it once frightened one of the guests Connie made me get rid of it. And now look what's happened!'

After the meal, we left the table and settled down in hammocks to talk. Tiny told us story after story as night fell. He spoke of his early days on the savannahs when there were so many jaguar around Karanambo that he had had to shoot one a fortnight in order to preserve his cattle. He remembered how a party of outlaws from Brazil used to cross the border on horse-stealing raids, until he went over to Brazil himself, held up the gang at pistol point, took away their guns and burnt down their houses. We listened fascinated. The frogs and crickets started calling; the bats fluttered in and out, and once a large toad wandered in and sat blinking owlishly in the light of the paraffin lamp slung from the roof.

'When I first came up here,' said Tiny, 'I hired a Macusi Indian to come and work for me. After I had given him an advance, I found out that he was a piaiman or witch doctor. If I had known that before, I wouldn't have hired him because witch doctors are never good workers. Soon after he had taken the money, he told me that he wasn't going to work any more. I said that if he tried to go away before he had worked off the money I had given him, I would beat him up. Well, he couldn't allow that to happen because he would lose face and then he wouldn't have any power among the other Macusi. I kept him until he had stayed long enough to clear off his advance and then I told him to go. When I did so, he told me that if I didn't pay him some more money he was going to blow on me, and if he did that, my eyes would turn to water and run out, I would get dysentery and all my bowels would drop out, and I would die. So I said "Go ahead and blow on me", and I just stood up and let him blow. When he had finished I said, "Well I don't know how Macusi blow, but I have lived a long time among the Akawaio and I am going to blow on you, Akawaio style." So I puffed myself up and jumped around him and blew. As I blew, I told him that his mouth would shut up and he wouldn't be able to eat anything; that he would bend backwards until his heels and his head touched and that then he would die! Well, I then dismissed him and I never thought any more about it. I went up into the mountains hunting, and it was some days before I returned. Soon after I arrived, my head Indian came in and said, "Massa Tiny, the man's dead!" I said, "There's plenty of people dead, boy. What man are you talking about?" "That man you blow upon, he's dead," he said. "When did he die?" I asked. "The day before yesterday. His mouth shut up the same as you said it would, he started to bend backwards and he died."

'And he was right,' said Tiny, concluding. 'The man *had* died, just as I had said he would.'

There was a long pause. 'But, Tiny,' I asked, 'there must be more to the story than that. It couldn't have been merely coincidence.'

'Well,' said Tiny, looking mildly at the ceiling, 'I had noticed a little sore on his foot and I knew that there had recently been two cases of tetanus in the village from which the man had come. Maybe that had something to do with it.'

—————

Sharing breakfast with the parrot and the hangnest, we discussed with Tiny our plans for the day. Jack had decided that he should unpack the cages, troughs and feeding bowls before starting to catch any animals.

Tiny turned to us. 'What about y'all, boys? Interested in some birds?' We nodded eagerly. 'Well, come along with me, I might be able to show you a few not far from here,' he said enigmatically.

Our walk with Tiny through the bush fringing the Rupununi River was an education in forest lore as, during the next half-hour, he pointed out to us a hole in a dead tree trunk trickling sawdust (the work of a carpenter bee), the spoor of an antelope, a magnificent purple orchid and the remains of an encampment where a party of Macusi had come to fish in the creeks. Soon he branched off the main path and cautioned us not to talk. The undergrowth was thicker and we tried to match his silent tread.

The vegetation here was festooned with a creeping grass which covered all the bushes with bright green loops and hung down in veils between them. Ignorantly and carelessly, I tried to brush some away with the back of my hand, but I quickly withdrew it in pain, for the creeper was razor grass, the stems and leaves of which are armed with rows of tiny sharp spines. My hand was cut and bleeding, and I said something louder than I should have done. Tiny turned round with his finger to his lips. Carefully picking our way through the tangle, we followed him. Soon the undergrowth became so thick that the easiest and most silent way of advancing was to wriggle forward on our stomachs, ducking under the razor grass.

At last he stopped and we drew alongside him. He carefully cut a small peephole in the thick blanket of razor grass which hung a few inches in front of our noses, and we peered through. In front of us lay a wide, swampy pond, its surface hidden by floating water hyacinth which here and there was in flower, so that the brilliant green carpet was splashed with small areas of delicate lilac-blue.

Fifteen yards beyond us the water hyacinth itself was obscured by the edge of an enormous flock of egrets which stretched across the centre of the lake and over to the other side.

'There you are, boys,' whispered Tiny. 'Any good to you?'
Charles and I nodded enthusiastically.

'Well, you won't want me,' Tiny continued. 'I'll get back for some breakfast. Good luck!' And he wriggled back soundlessly, leaving the two of us alone peeping through the razor grass. We looked again at the egrets. Two species were mingled in the flock; great egrets and the smaller snowy egrets. Through binoculars, we could see them raising their delicate filigree crests as they squabbled among themselves. Occasionally a couple would rise vertically in the air, sparring frenziedly with their beaks, only to subside as suddenly as they had risen.

Towards the far edge of the lake, we could see several tall jabiru storks standing head and shoulders above the other birds, their black naked heads and scarlet dropsical necks standing out vividly amid the pure white of the egrets. In the shallows on the far left, there were hundreds of ducks. Some were lined up in pert regiments, each one facing the same way with military precision, others floated in squadrons on the pond itself. Close to us, a lily-trotter or jacana trod cautiously on the floating leaves of the water hyacinth, its weight spread over several plants by its enormously elongated toes which made it lift its feet at each step with the action of a man in snowshoes.

Loveliest of all, within a few yards of us we saw four roseate spoonbills. As they dabbled busily in the shallow water, sifting the mud through their bills in search of small animal food, they looked ravishingly beautiful, for their feathers were suffused with the most delicate shades of pink. But every few minutes they lifted their heads to gaze around, and we saw that the ends of their bills were enlarged into flat discs which gave them a slightly comic look, oddly at variance with the grace and beauty of their bodies.

We set up the camera to begin filming this magnificent scene, but no matter where we placed it a small isolated bush in front of us impeded our view. We held a whispered council and decided to risk scaring the birds and advance across a few yards of lush grass to a spot underneath the bush which seemed just large enough to accommodate us both with the camera. If only we could reach it without causing alarm we should have a clear, uninterrupted view of all the birds on the lake – ducks, egrets, storks and spoonbills.

As quietly as possible we enlarged our peephole in the veil of razor grass into a slit. Pushing the camera in front of us, we slowly wriggled out and across the grass. Charles gained the bush safely and I joined him. In slow motion, lest a sudden movement should scare the birds, we erected the tripod and screwed the camera into position. Charles had almost focused on the spoonbills, when I put my hand on his arm.

'Look over there,' I whispered, and pointed to the far left of the lake. Sloshing their way through the shallows came a herd of savannah cattle. My immediate concern was that they might scare the spoonbills just as we were in a position to film them, but the birds took no notice. The cows came ponderously towards us, swinging their heads. In front of them walked a single leader cow. She stopped, lifted her head and snuffed the air. The rest of the herd stopped behind her. Then she advanced purposefully towards our little bush. When she was some fifteen yards away she stopped again, let out a bellow, and pawed the ground. From where we lay, she looked a very different animal from the gentle Guernseys of an English pasture. She bellowed again impatiently and brandished her horns at us. I felt very vulnerable lying there; if she charged she would come over the bush like a steamroller.

'If she charges,' I whispered to Charles nervously, 'she'll scare the birds, you know.'

'She might also damage the camera and then we should be in a mess,' whispered Charles.

'I think perhaps it might be wiser to retreat, don't you?' I said, with my eyes fixed on the cow. But Charles was already on his way, wriggling back to our razor grass thicket and pushing his camera in front of him.

We sat well back in the bushes and felt foolish. To have come all the way to South America, the home of jaguar, venomous snakes and cannibal fish, and then to be frightened by a cow, seemed a little ignominious. We lit cigarettes and persuaded ourselves that for once discretion had indeed been the better part of valour, if only for the sake of our equipment.

After ten minutes, we decided to see if the cows were still there. They were, but they took no notice of us as we lay in our thicket. Then Charles pointed to a wisp of grass in front of us swaying gently in the breeze away from the cattle. The wind had changed and it was now in our favour. Emboldened by this, we once more wriggled out to the small bush and set up the cameras. For two hours we lay there, filming the egrets and the spoonbills. We watched and recorded a little drama in which two vultures found the head of a fish on the margin of the lake, only to be driven from their booty by an eagle, which then became so nervous of a counter-raid by the vultures that it could not settle down to eat the head and finally had to fly away with it. An hour before we had finished, the cows splashed their way back to the savannahs.

'What a wonderful sight it would be if all these birds took to flight,' I whispered to Charles. 'Edge your way out of the bush; I'll leap out on the other side, then as they take off stand up and film them wheeling against the sky.' With great care and moving very slowly so as not to startle the flock prematurely, Charles crept from under the bush and crouched by its side clutching his cameras.

'Right! Stand by!' I whispered melodramatically and with a shout I leapt from the bush waving my arms. The egrets took not the slightest notice. I clapped and shouted and there was still no movement. This was absurd. All that morning we had crept with infinite stealth through the bush, hardly daring to whisper lest we should frighten these supposedly timid birds, and now here we were standing shouting at the tops of our voices, yet the entire flock appeared totally unconcerned and our silence seemed to have been quite unwarranted. I laughed out loud and ran towards the edge of the lake. At last the ducks nearest to me took off. The egrets followed them and in a great surge the whole white flock peeled from the surface of the lake and swept into the air, their calls echoing over the rippling water.

Back at Karanambo, we confessed to Tiny our fear of cows.

'Well,' he laughed, 'they do get a bit skittish sometimes and I have had to run for it myself before now.' We felt our reputations had not yet been irretrievably lost.

———

The next day Tiny took us to a stretch of the Rupununi River just below his house. As we walked along the banks, he pointed out to us a series of deep potholes which riddled the soft tufa-like rock. He dropped a stone down one and an asthmatic belching echoed up from the pool in the bottom of the hole.

'One's at home,' said Tiny. 'There's an electric eel living in almost every one of these holes.'

But I had another means of detecting the eels. Before we left England, we had been asked to record the electric impulses of these fish on our tape recorder. The apparatus needed was simple – two small copper rods fixed in a piece of wood about six inches apart and connected to a length of flex which could be plugged into our machine. I lowered this elementary piece of equipment into the hole and imme-diately heard on my small earphones the electric discharge of the eel recorded as a series of clicks, which increased in volume and frequency, rose to a climax and then subsided. This discharge is thought to act as a type of direction-finding device, for the eel possesses special sensitive organs all along its lateral line which enable it to detect the changes in elec-tric potential caused by solid bodies in the water, and so to solve the problem of manoeuvring its six-foot length among the rocks and crannies in the murky depths of the river. In addition to this minor semi-continuous discharge, the eel is also capable of delivering an immense high voltage shock with which it is supposed to kill its prey and which, it is said, is powerful enough to stun a man.

We moved on down to Tiny's landing and climbed into two canoes, powered by outboard motors which drove us steadily upriver, passing on our way a tree colonized by a number of hangnest birds, their nests dangling like giant clubs from the branches. Behind us we trailed handlines baited with spinning metallic lures in the hope that we might catch some fish. Almost immediately I had a bite. Hauling in my line, I found a silvery-black fish, twelve inches long, and began removing the hook from its mouth.

'Watch your fingers,' Tiny remarked idly. 'That's a cannibal fish you've got there.'

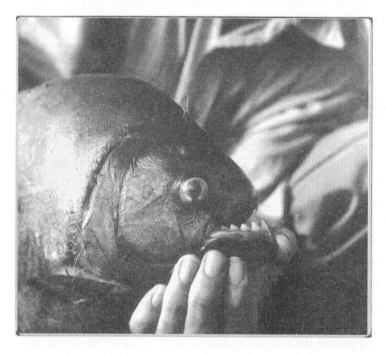

A piranha

I dropped it hastily on the bottom of the boat.

'Don't do that, man,' said Tiny, a little aggrieved, seizing a paddle and giving the creature a clout which stunned it. 'He might give you a nasty nip.' He picked up the fish, and to prove his point stuck a piece of bamboo in its open mouth. The rows of triangular razor-edged teeth clashed shut on the bamboo, cutting it as cleanly as an axe-blow.

I watched appalled. 'Is it really true that if a man fell in a shoal of those things, he would be hauled out a skeleton?' I asked.

Tiny laughed. 'Well, I reckon that piranha, or perai as we call them, might make quite a mess of you, if you were silly enough to stay in the water once they started biting. It's usually the taste of blood that makes them attack in the first place, so I shouldn't go bathing if you've got an open cut. Luckily they don't like broken water, and so when you get out of a canoe and haul it up rapids you needn't worry, they are seldom there.

'Of course,' he went on, 'they sometimes attack unprovoked. I remember once getting into a canoe with fifteen Indians. We got in one at a time, and in doing so we had to put one foot in the water. No one except me had boots on. I was last in, and as I sat down, I noticed that the Indian in front of me was bleeding badly. I asked him if he was all right and he said that a perai had bitten him as he got in. It turned out that thirteen out of the fifteen had small pieces of flesh bitten clean out of their feet. None of them had cried out at the time, and no one had thought of warning the men coming afterwards. Still, I suppose that story tells you more about Indians than it does about perai.'

After several days at Karanambo, we returned to Lethem. Slowly the animal collection grew and when, after two weeks on the savannahs, we flew back to Georgetown, we took with us not only our caiman, lying in a huge tailor-made wooden crate, but a giant anteater, a small anaconda, some fresh water turtles, capuchin monkeys, parakeets and macaws. It seemed a reasonable beginning.

Charles Lagus flying back from the Rupununi

3

The Painted Cliff

The River Mazaruni rises in the highlands of the far west of Guyana, close to the Venezuelan border. For a hundred miles it winds round three parts of a huge circle before it breaks through the girdle of high sandstone mountains which enclose it and, over the short distance of twenty miles, descends thirteen hundred feet in a series of cascades and rapids which form an impassable barrier to river traffic.

The only land routes into the basin are long and arduous trails over the mountains, the easiest of them involving a three-day march through thick, difficult forest and a climb over a three-thousand-foot pass. The whole area, therefore, was virtually cut off from the rest of the country, and the fifteen hundred Amerindians who lived there had, until a few years before our visit, remained isolated and relatively untouched by the civilization of the coast.

But the arrival of the aeroplane in the country had completely changed the situation, for by amphibian plane it became possible to fly over the mountain barrier and land in the centre of the basin on a long, wide stretch of the Mazaruni River. This sudden accessibility might have had serious consequences for the Akawaio and Arecuna tribes living there, so to prevent their possible exploitation, the Government had declared the whole area an Amerindian reserve – forbidden country for diamond and gold prospectors and for travellers without permits. It had also appointed a District Officer whose job it was to watch over the welfare of the Amerindians.

Bill Seggar held that post, and when we first arrived in the country he was, fortunately for us, paying one of his infrequent visits to Georgetown to buy six months' supply of food, trade goods, petrol and other necessities which had to be flown in to his station.

He was a tall, dark, heavily built man, with a deeply lined face. Rather laconically, lest he should betray too much of the enthusiasm and pride which he felt for his province, he had told us of its wonders; of newly discovered waterfalls, of huge areas of unexplored forests, of the strange 'hallelujah' religion of the Akawaios, of hummingbirds, tapirs and macaws. He had estimated that he would have finished his business in Georgetown by the time we came back from our fortnight's visit to the Rupununi and had generously suggested that we might fly back with him to the basin.

So it was with great excitement that we now looked for Bill in Georgetown, to discover when his plane was leaving. We eventually ran him to earth in the bar of a hotel, gloomily staring at a glass of rum and ginger. He had bad news. The stores he had ordered were due to be flown in to the area by Dakota aircraft which normally landed on a small patch of open savannah near the eastern margin of the basin at Imbaimadai. This strip is usually serviceable throughout the long dry season, but during the rains it becomes waterlogged and useless. Theoretically, we should be able to use it now, in mid-April, but there had been a freakish outburst of rains which had converted the airstrip into a quagmire. Bill was going to fly into the area the next day by amphibian plane, land on the Mazaruni River just below the savannah at Imbaimadai, and then squat on the airfield, reporting its condition by radio day by day, so that as soon as it had dried out the freight plane could take off from Georgetown to bring in the essential stores. These supplies obviously must come first, but if they got in safely and if the strip was still dry, then we could follow as a last load. We finished our drinks moodily and said goodbye to Bill, wishing him luck when he took off next morning for Imbaimadai.

We waited in Georgetown, anxiously visiting the Department of the Interior each day for news of the airstrip. On the second day, we heard that the rain had stopped and that, given sun and no more rain, the airstrip should dry out and be serviceable in about four days. We spent those four days helping Tim Vinall settle the animals we had caught on the Rupununi in comfort-able living quarters. The Agricultural Department had lent us a garage in the Botanic Gardens, which we quickly converted into a miniature zoo with cages stacked in tiers around its walls. Some of the bigger animals could not be accommodated, and very generously the Georgetown Zoo offered to take several of them, including the giant anteater, as temporary boarders. The caiman in its crate lay half-submerged in one of the canals in the gardens.

At the end of the four days, a radio message was received from Bill Seggar saying that all was well and that the freight plane could leave. All that day and the next, stores were ferried in to him. Then at last it was our turn.

We said our farewells to Tim, whose unenviable job it was to remain in Georgetown looking after the Rupununi animals, and once more with all our equipment we climbed into a Dakota aircraft.

Flying over rain forest is rather boring. Beneath us stretched an unbounded, featureless ocean of green. The myriad exciting forms of animal life which we knew it contained lay concealed beneath its dimpled green surface, though occasionally birds skimmed like flying fish above the crests of the trees. Once in a while we saw little clearings, dotted with tiny huts, like islands in the sea of forest.

After an hour, however, the prospect changed, for we were approaching the Pakaraima Mountains, which form the south-eastern part of the Mazaruni's mountainous defences. The forests climbed on their flanks until, here and there, the slopes became so steep that no trees could grow upon them, and the mountain-side became a naked precipice of cream-coloured rock.

In a few minutes we sailed over these barriers which had proved so formidable to early travellers, and there below us wound the young Mazaruni River which even here was some fifty yards wide. Then, as if by a miracle, we saw below us in the middle of the forest a small patch of open savannah and towards one side of it a hut and two tiny white figures, which we knew were Bill and Daphne Seggar.

The Dakota circled and came down for a landing. Through no fault of the pilot's, it was a rather bumpy one, for the Imbaimadai airstrip had no tarmac runways; it was simply open country from which the larger boulders and the more obvious trees and bushes had been removed by Bill Seggar's Amerindian helpers.

The Seggars walked up to greet us. Both of them were bare-foot, she tall and lithe in an athlete's woollen tracksuit, he in a pair of khaki shorts and a shirt open to the waist, his hair still wet from a bathe in the river. Bill was highly relieved to see us, for with us in the plane we had the last of his essential stores; now, come what may, his provisions would last him through the rainy season. This he anticipated would not start for at least another month, and, all being well, we should be able to leave from the Imbaimadai airstrip four weeks hence.

'But,' he said, 'you can never tell. The rains may start again tomorrow. If they do, though,' he added cheerfully, 'we shall always be able to ship you out in instalments by amphibian plane at phenomenal cost.'

We spent the night in the semi-ruined hut on the Imbaimadai airstrip, and the next morning Bill suggested that we should push on to the head waters of the Mazaruni and travel up the Karowrieng, one of the smaller tributaries of the main river, into uninhabited and relatively unexplored country. We asked what we might see.

'Well,' said Bill, 'nobody lives up there, so there must be plenty of wild life to interest you. There is also a pleasant waterfall, which I discovered a year or two ago, and some mysterious

Amerindian cliff paintings which few people have ever seen and which no one seems to know much about. You might take a look at them too.'

Bill was expecting further plane-loads of goods, though these were not as vital as those he already had. The first load, however, was not due to arrive for another two days, and the next morning he suggested that he and Daphne should accompany us on the first day of our journey. Accordingly, the five of us climbed into a huge forty-foot dugout canoe fitted with a powerful outboard engine, which Bill habitually used in travelling about his area. A crew of six Amerindian boys came with us.

That day was a fascinating one for us; it was the first time we had had a close view of the forest. We travelled along a canyon of sunlight; beneath us the placid, translucent brown river, and on either side of us the vertical green walls of the forest. Purple-heart, green-heart and mora trees grew on the banks to a height of a hundred and fifty feet. Beneath their crowns, matted creepers and lianas dangled in a curtain which screened us from the interior of the forest. Nearer the ground, smaller bushes reached greedily outwards for the sunlight which was denied them in the gloomy depths of the forest. This leafy façade was not a uniform green, for as the rainy season was approaching some of the trees were sprouting fresh growths of amber-red leaves which hung limply downwards, forming clouds of vertical lines strikingly prominent among the riotous exuberance of the rest of the vegetation.

Two hours' journey brought us to a series of rapids. The river here tumbled over a wide barrier of rocks, which churned its amber brown into a creamy white. We unloaded the most delicate and easily damaged pieces of our equipment – cameras and recording machines – and carried them across a portage to the top of the rapids, and returned to help the Amerindians drag the heavy canoe over the rocks. It was a hot and tiring job, but the men laughed over it and were convulsed with mirth when one of us clumsily lost a foothold and fell up to the waist in an unexpectedly deep cleft between the boulders. At last we hauled the canoe into the still, black pool which marked the top of the rapids, and once more we were on our way.

Hauling up the rapids

After another hour's travel, Bill told us to listen; above the noise of our engine we could hear a distant boom.

'My waterfall,' he said.

Fifteen minutes more brought us to a bend in the river. The sound of the waterfall was now very loud indeed and Bill told us that it lay just round the curve. To go further upriver would involve an arduous portage of the canoe round the falls and so we decided to make camp for the night on the bank. Bill and Daphne, however, could not stay with us, for they had to return to Imbaimadai to bring in the remaining plane-loads of stores.

Before they left, and while the Amerindians were clearing a camp site, they walked with us along the river bank to look at the waterfall. Guyana is rich in waterfalls. Only a few miles away south were the eight-hundred-foot Kaiteur Falls, so that in Guyanese terms, Bill's falls were negligible — a mere hundred feet high; yet as we rounded the bend in the river, they were a startlingly beautiful sight. A sickle-shaped, white sheet of foam thundered over an overhanging ledge and fell sheer into a wide, open pool at the base. We swam in the pool; we climbed among the tumbled boulders at the fall's base and scrambled round to the dank cavern underneath the falls, through which swifts were flitting.

Jack Lester at Maipuri Falls

Bill had christened his falls the Maipuri — after the local name for tapir — the tracks of which he had found on the river banks when he had first discovered them. Unfortunately we could not spend much time sightseeing, for if Bill and Daphne were to get back to Imbaimadai before dark, they would have to return almost at once, so we retraced our steps to the Amerindians and the canoe.

Bill and Daphne, taking two of the men with them, set off again downriver, leaving us with the promise that they would send the canoe back with an Amerindian in two days' time to collect us.

They had left us with four Amerindians to help carry our gear wherever we might wish to go in the forest. They were all Akawaios, but as they all worked on Bill's station, they were partly Europeanized and wore khaki shorts and shirts and spoke a pidgin English, a dialect, spoken and understood by Guyanese of all kinds – Amerindian, Afro-Caribbean, East Indian and European. Though based on a simplified form of English it, like most pidgins elsewhere in the world, has its own particular rules, vocabulary, simplifications and pronunciations. Verbs for the most part are avoided, or if used at all only in the present tense, and plurals and emphases are conveyed by simply repeating a word several times. So we spoke back in pidgin too, and understood each other perfectly well. The senior man, Kenneth, also understood some, if not all, of the intricacies of the outboard engine, though we were to discover that his main method of dealing with any fault in the engine was to remove all its plugs and blow down them. His first lieutenant was named King George – a stocky shock-headed man with a permanent ferocious scowl. We understood from Bill that he was a headman of a village farther down the river and had adopted this royal title himself. Efforts had been made to try to make him change his name to George King, but he had stoutly refused to do so.

While we had been looking at the falls, the four Akawaios had cleared a large area in the bush some fifteen yards square and had built a framework from saplings cut in the forest and bound together with pieces of bark and liana, over which they had stretched a large tarpaulin to protect us from a sudden rain-storm. Underneath this we were to sling our hammocks. A fire was already burning and water boiling. Kenneth came up to us with a gun in his hand and asked what sort of bird we would like for our supper. We suggested maam, the lesser tinamou, a small flightless bird rather like a partridge, which makes very good eating.

'Very well, sir,' said Kenneth confidently, and disappeared into the forest.

An hour later, he returned with a large, fat maam, just as he had promised. I asked how he was able to find just the bird for which we had asked, and he told me that all Amerindians hunt by imitating bird calls. We had decided on maam, so he had gone into the forest and, moving stealthily, imitated the call of the maam, which is a long low whistle. After thirty minutes a bird had replied. Calling continuously, he had crept closer and closer and had finally shot it.

After supper, we climbed into our hammocks and settled down for our first night in the forest. Our fortnight on the savannahs had taught us something of the technique of hammock sleeping, but on the savannahs it had been hot throughout the nights as well as the day but here, high in the Mazaruni Basin, the nights were very cold. I learnt that night that a hammock sleeper should take with him twice as many blankets as he would normally use in a bed, for he has to wrap himself below as well as above and the efficiency of a single blanket is thereby divided by two. It was so cold that, after an hour, I had to climb out of my hammock and put on all the spare clothes I had brought with me before I could get to sleep; even so, I passed a bad night.

I woke well before dawn, but as the sun rose I was amply rewarded, for the calls of macaws and parrots echoed over the river and a hummingbird was already feeding on the blossoms of a creeper which hung down by the water's edge. It was a tiny, bejewelled creature, no bigger than a walnut, moving jerkily through the air. When it decided to feed from a flower, it hovered in front of it, flashing out its long, threadlike tongue and sipping the nectar from the depths of the flower. When it finished, it slowly reversed through the air on its rapidly beating wings and shot away in search of another blossom.

After breakfast, King George told us that the paintings Bill Seggar had mentioned lay two hours' march away in the forest.

We asked him if he could lead us to them. He said that he had only been there once before, but that he was sure that he could find them again. With another of the Akawaios to help carry our cameras, he led us off into the bush. He went ahead unhesitatingly, cutting notches in the trees and bending over the heads of saplings to mark the route, so that we should not lose our way back. We were now in high, tropical rain forest. The great trees rose two hundred feet above us, most of them covered in plants which have the peculiar habit of not growing in earth, but of sending down long, aerial roots to draw nourishment from the humid air. Occasionally, on the forest floor, we came across a wide area scattered thickly with fallen, yellow blossoms, which made a carpet of colour in the gloomy forest. We looked up to see where they had come from, but all the trees rose so high above us that if it had not been for the fallen flowers we might never have guessed that any of them flowered at all.

In between the boles of the trees, there was a tangle of small saplings and creepers through which we had to cut a passage with our knives. We never saw a large animal, but we were well aware of the presence of innumerable tiny creatures around us, for the air was filled with the chirps and pipings of frogs, crickets and other insects.

After two hours' hard going, both Charles and I were very tired indeed. It was hot and muggy and we were soaked in sweat and very thirsty. We had not seen any water to drink since we left the river.

And then, suddenly, we came upon the cliff for which we had been searching. It rose vertically for several hundred feet and broke through the forest canopy, the shade of which had hitherto kept us in sweltering twilight. The rock and branches did not meet, and through the gap between them a shaft of sunlight struck diagonally down on to the white quartzite rock of the cliff, floodlighting the red and black paintings with which the rock was smothered. The sight was so impressive and so

THE PAINTED CLIFF

startling that weariness dropped from us and we raced excitedly
to the foot of the cliff.

The paintings stretched for forty or fifty yards along its base
and rose to a height of thirty or forty feet. The designs were
crude, but many of them clearly represented animals. There
were several groups of birds, probably maam, which Kenneth
had hunted for us the previous evening, and many indeterminate quadrupeds. One seemed to us to be an armadillo, but
then if we regarded the armadillo's head as a tail, the design
became equally clearly a representation of an anteater. Another
creature lay upside down, its feet in the air. At first we thought
that this might represent a dead beast, but then we saw that it
had two claws on its forelegs and three on its hind, the number
possessed by a two-toed sloth. Above it, to corroborate our
identification, was a thick, red line, obviously the branch on
which the sloth should hang, but which, perhaps, presented
difficulties in drawing to the unknown artist, who therefore
painted it separately above the creature, to make his meaning
clear. Among the animals were boldly painted symbols: squares,
zigzags, and strings of lozenges, the meaning of which we could
not begin to guess at.

Most moving and evocative of all, interspersed between the
animals and symbols, were hundreds of handprints. On the higher
parts of the cliff they were in groups of six or eight, but near
the base they were so numerous that they had been superimposed
one upon the other to form almost solid areas of red paint. I
placed my hand over several of them and found them all to be
smaller than mine. At my request, King George made the same
comparison; the prints fitted his hands exactly.

I asked King George if he could tell us what the paintings
represented, but though he willingly made several wild suggestions for each animal we pointed to, his identifications were
obviously just as tentative as ours. If we made alternative suggestions, he agreed and laughed and confessed he did not know;
but one design we all agreed upon. 'What's that?' I asked him,

39

Handprints on the cliff wall

Animal designs, perhaps a sloth and a giant anteater

pointing to the outline of an upright, human figure which was very obviously male. King George was convulsed with laughter.

'He sporting,' he said with a wide grin.

King George was most emphatic that he knew neither the significance nor the origin of the paintings. 'They made long, long time ago,' he explained, 'but not by Akawaio man.' We found evidence of their antiquity, for here and there part of the hard rock had flaked away, taking with it a section of the paintings; the resultant scars were no longer fresh, but had weathered to the same shade as the rest of the cliff, a process which must have taken a great many years.

Their purpose, whatever it was, must have been an important one, for to place the designs so high on the cliffs, the artists must have gone to the labour of building special ladders. Perhaps the paintings were part of a magical ceremony connected with hunting, the man drawing the animal he desired and then registering his identity by leaving his handprint. Yet only one of the creatures, a bird, was shown as being dead and none appeared as being wounded, as they seem to be in the Palaeolithic painted caves of France. For an hour Charles and I photographed the designs, building a crude ladder from saplings to enable us to reach the higher ones.

My thirst became overpowering, and as I came down the ladder for the last time, I noticed that water was dripping from the top of the overhanging cliff on to a boulder which was covered in thick, sodden moss. I hurried to the spot, squeezed lumps of the moss and moistened my mouth with the gritty, dark-brown water. Seeing me do this, King George disappeared among the cliffs to the left and within five minutes returned to say that he had discovered water. I followed him, clambering over the huge boulders that littered the base of the cliff. A hundred yards to the left, a wide crack ran down the cliff face. As it approached the ground, it broadened and deepened into a small cave, the floor of which was formed by a deep black pool of water. In the back of the grotto, a vigorous stream of water tumbled into the pool; but the pool had no visible outlet. It presented such a startling appearance

– this torrent spouting from the living rock and pouring into a seemingly bottomless pool which never overflowed – that for the moment I forgot my thirst. Surely such a pool could, to primitive minds, invest the cliff with a magical character. I remembered the grottoes of ancient Greece, into which sacrificial objects had been thrown to appease the gods, and plunged my arm into the water in the hopes of finding a stone axehead; but the pool was so deep that I could only touch the bottom in the shallower part and there I found only gravel. I tested the pool with a stick and found it to be over five feet deep.

The spring at the bottom of the cliff

My thirst quenched, I returned to the cliffs to tell Charles of the discovery. We sat and speculated as to what the paintings might mean and whether the grotto had any connection with them. By now the sun had disappeared over the crest of the cliff and the paintings had lost their theatrical lighting. If we were to regain camp that night we should have to start back immediately.

4

Sloths and Snakes

W e could have spent much time wandering in the forests and yet seeing little, so we decided to augment our limited knowledge and experience by recruiting two of the Akawaios who worked on the station to accompany us on our jaunts. Their eyes were more skilled than ours in spotting the smaller animals and they also knew the forest so intimately that they could take us to flowering trees that might be attracting hummingbirds and to others in fruit that might be visited by flocks of parrots or troops of monkeys.

Our first major success, however, was scored by Jack. We were walking through the forest not far from the airstrip, picking our way through spiny creepers. We paused at the base of one of the largest trees we had so far found. From its branches high above us hung thick lianas in immobile contortions. If we could have concentrated several years of the lianas' movements into a few minutes, we should have seen them twisting and writhing, strangling both themselves and the trees from which they hung. Jack looked up into the tangle.

'Is there anything up there, or is it my imagination?' he said softly.

I could see nothing. Jack explained more carefully where I should look and at last I saw what he had spotted – a round grey shape hanging upside down from a liana. It was a sloth.

Sloths are incapable of rapid movement and, for a change, there was no risk that this animal would career off and be lost in a few seconds in the higher reaches of the forest roof. There was enough time for us to decide that Charles should film the capture,

43

that Jack's ribs were still painful from a recent fall to prevent him from doing anything strenuous, and that I, therefore, should be the one to climb up and bring the mysterious creature down.

The ascent was not difficult, for the dangling creepers provided an abundance of holds. The sloth saw me coming and in a slow-motion frenzy began climbing hand-over-hand up its liana. It moved so slowly that I was able to overhaul it with ease, and forty feet above the ground I caught up with it.

Charles Lagus filming in the forest

The sloth, about the size of a large sheepdog, hung upside down and stared at me with an expression of ineffable sadness on its furry face. Slowly it opened its mouth, exposing its black enamel-less teeth, and did its best to frighten me by making the loudest noise of which it is capable – a faint bronchial wheeze. I stretched out my hand and, in reply, the creature made a slow, ponderous swing at me with its foreleg. I drew back and it blinked mildly, as if surprised that it had failed to hook me.

Its two attempts at active defence having been unsuccessful, it now concentrated on clinging firmly to the liana. Loosening its grip was not easy, for my own position was somewhat precarious. Holding on to my own liana with one hand, I reached over with the other and tried to detach the sloth. As I prised loose the scimitar-sharp claws on one foot and began work on the next, the sloth, very sensibly and with maddening deliberation, replaced its loosened foot. At no time did I manage to get more than one limb free at a time. I continued for five minutes in this way, not substantially helped by the ribald suggestions that Jack and Charles shouted up to me. Plainly, this one-handed struggle could go on for ever.

Then I had an idea: close by me hung a thin, crinkled liana, nicknamed by the Akawaios 'granny's backbone'. I called down to Jack and asked him to cut it loose near the ground. I then pulled the severed end up to me and dangled it near the sloth as I unfastened each of its legs. The animal was so determined to grasp anything within reach that, limb by limb, I was able to transfer it to the smaller liana. That done, I gently lowered the liana so that the sloth, clinging obligingly to the end, slowly descended straight into Jack's arms. I clambered down.

'Nice, isn't it?' I said. 'And it's a different species from the one I remember seeing in the Zoo.'

'Yes, it is,' Jack replied mournfully. 'The one in London is a two-toed sloth. It has been there for several years, feeding quite happily on apples, lettuce and carrots. This one here is the three-toed species. You've never seen it in London for the simple reason that it will only eat cecropia plant, and while there's plenty of cecropia in the forest here, there's none to be got in London.'

We knew therefore that we had to release it, but before doing so, we decided to keep it for a few days so that we could watch and film it. We carried it back and put it on the ground near the base of an isolated mango tree, near the house. Without a branch to hang from, the sloth had the greatest difficulty in moving at all. Its long legs splayed out and it was only by laboriously humping its body that it managed to drag itself across the few yards that separated it from the bole of the mango tree. Once the creature was there, however, it clambered gracefully up the trunk and contentedly suspended itself beneath one of the boughs.

Every feature of its body seemed to have been modified in some way to suit its inverted existence. Its grey, shaggy hair, instead of flowing down from its backbone towards its stomach as in any normal creature, was parted along its belly and flowed towards its spine. Its feet were so extensively adapted to act as hangers that they had lost all sign of a palm and the hook-like claws appeared to project straight from a furry stump.

A wide circle of vision is obviously very necessary when hanging in the tree-tops, and the animal had a long neck which enabled it to twist its head through almost a full circle. The sloth's neck bones are of considerable interest to the biologist, for whereas nearly all mammals, from mice to giraffes, have only seven bones in their neck, the three-toed sloth has nine. It is tempting to conclude that this also is a special adaptation for an upside down life. Unfortunately for the theorists, however, the two-toed sloth, which lives in exactly the same manner, and which can perform similar feats of neck mobility, has only six neck bones, one fewer than nearly all other mammals.

On the third day, we noticed our sloth craning forward in an endeavour to lick something on its hip. Curious, we looked closer, and to our astonishment saw that it was caressing a tiny baby, still wet, that must have been born only a few minutes earlier.

The fur of a sloth is supposed to support a growth of micro-scopic plants, giving the creature a greenish-brown tinge which is of considerable value to it as camouflage. The birth of this baby, however, did not corroborate this, for the infant could not yet have accumulated its own garden in its coat, and yet it was exactly the same colour as its mother. Indeed, when it had dried we had the greatest difficulty in distinguishing it as it nestled in its mother's shaggy fur, occasionally groping along the length of her enormous body to suck from the nipples in her armpits.

The three-toed sloth with its baby

We watched the pair for two days, the mother tenderly licking her baby, sometimes detaching one of her legs from the bough above her to support her tiny offspring. The birth seemed to have robbed her of her appetite and she no longer took slow bites from the cecropia which we tied to her tree. Rather than run the risk of her going hungry, we carried the two back to the forest. There we hooked her on to a liana, and with her baby peering at us over her shoulder she started to climb upwards.

When we returned to the spot an hour later to make sure that all was well, mother and child were nowhere to be seen.

Soon after we released the sloth, Bill and Daphne Seggar had to leave us, their canoe piled high with stores. There still remained a large dump of supplies on the airstrip at Imbaimadai and Kenneth was to return the next day in the canoe to bring down another load. Jack had decided that he should remain at our base for another few days and concentrate on collecting animals in the immediate neighbourhood. But part of our filming schedule was to record something of the everyday life in an Amerindian village and Bill suggested that, as engine fuel was at a premium, Charles and I should travel upriver with Kenneth, land at one of the villages, and stay there for a time.

'I should go to Wailamepu first,' Bill said. 'It's a short way up the Kako River, a tributary of the Mazaruni. One of the villagers is a bright young lad named Clarence who once worked for me here and who consequently speaks quite good English.'

'Clarence?' I asked. 'That seems rather an odd name for an Akawaio.'

'Well, the Indians used to believe in "Hallelujah", an odd version of Christianity which arose in the southern part of Guyana in the early nineteenth century; but Seventh Day Adventist missionaries converted the inhabitants of Wailamepu village and in the process rechristened them with European names.

'Of course,' he went on, 'the old names are still used among themselves, but I don't think you will find many Indians who will tell you their Akawaio name.' He laughed. 'They seem to be able to combine their old beliefs quite conveniently with the new ones taught by the missionaries and will shift from one to the other whenever it suits them.

'The Adventists teach, for example, that you should not eat rabbit. Of course, there are no rabbits here but a large rodent called a labba is roughly equivalent. Unfortunately, labba-meat has always been one of the favourite foods of the Indians and forbidding it was quite a blow to them. There is a story that a missionary once came across one of his Indian converts cooking a labba over a fire. He told the Indian how sinful it was.

'"But this not labba," the Indian said, "this fish." "No fish has two big front teeth like that," replied the missionary crossly, "you speak nonsense." "No, sir!" said the Indian. "You know how, when you first came this village, you say my Indian name bad name, and you sprinkled water over me and say my name now John. Well, sir, I walk in the forest today, I see labba and I shoot 'im, and before he die, I throw water over him and I say 'Labba is bad name, you now fish.' And so now I eat fish, sir.'"

———

The next morning we set off for Wailamepu with Kenneth and King George. The outboard engine was working perfectly and in two hours we reached the mouth of the Kako River. After fifteen minutes' journey up the Kako, we saw a path running up the river bank into the forest. At its foot, by a muddy landing, were moored several canoes. We stopped our engine, disembarked and walked up the path to the village.

Scattered around a sandy clearing were eight rectangular ridge-roofed huts, raised on short stilts. Their walls and floors were of bark, and they were thatched with palm leaves. Women, some in worn cotton frocks, others wearing only the traditional bead apron around their loins, stood at the doors of the houses and watched us. Scraggy chickens and mangy dogs wandered in and out of the huts and tiny lizards skittered from beneath our feet as we went.

Kenneth led up us to an old, genial man sitting in the sun on the steps of his house. He was naked except for a pair of tattered, heavily patched shorts which had once been khaki.

'This, headman,' said Kenneth, and he introduced us. The headman spoke no English, but through Kenneth he welcomed us and suggested that we should stay in a derelict hut at the far end of the village, which was used as a church when the missionary came to the district. Meanwhile, King George had enlisted the aid of some of the village boys, and between them they had brought all our baggage from the boat and laid it in a pile near the church.

We walked back to the river with Kenneth and King George. Kenneth wrestled with the outboard engine. At last it started and the boat surged away from the bank. 'I come back one week's time,' yelled Kenneth above the roar of the engine as he disappeared downriver.

Most of that day we spent in unpacking our gear and constructing a little kitchen outside the hut. Later we wandered about the village endeavouring not to appear too inquisitive too early, for it seemed hardly polite to start peering into huts and taking photographs until we had got to know the villagers. We soon discovered Clarence, a cheerful man in his early twenties, sitting in his hammock, busy weaving an elaborate piece of basketwork. He welcomed us with genuine sincerity but made it clear that at that moment he was too busy to do more than exchange a few words.

We returned to the church in the late afternoon and began thinking of making a meal.

Clarence appeared at the door.

'Good night,' he said with an expansive smile.

'Good night,' we replied, having been forewarned that this was the normal evening greeting.

'I bring you these t'ings,' he said, putting down three large pineapples on the floor. He sat down and settled himself comfortably in the doorway, his back against the doorpost.

'You come from long way?' We admitted that we did.

'An' why you come here?'

'Our people, far far away across the sea, do not know anything about the Akawaio on the Mazaruni. We bring all kinds of machines to take pictures and sounds, so that we can show our people how you make cassava bread and woodskin canoes and all things like that.'

Clarence looked incredulous.

'You t'ink anyone far away wish to know those t'ings?'

'Yes, for sure.'

'Well, the people here will show you if you really wish,' said Clarence, still slightly doubtful. 'But please, you show me all these t'ings you bring.'

Charles produced the camera and Clarence peered down the viewfinder with delight. I demonstrated the tape recorder. This was an even greater success.

'These fine t'ings,' said Clarence, his eyes gleaming with enthusiasm.

'There is one other thing we come for,' I said. 'We wish to find all kinds of animals: birds, snakes, every kind of thing we want.'

'Ah ha!' said Clarence. 'King George told me 'bout one man you leave bottom-side at Kamarang, who able catch snakes and get no fear at all. That true t'ing that King George say?'

'Yes,' I replied. 'My friend he catch all things.'

'You catch snakes too?' inquired Clarence.

'Well, yes,' I replied modestly, anxious not to let pass this chance of gratuitous prestige.

Clarence pressed the matter.

'Even the kind that bite bad?'

'Er – yes,' I said, rather uncomfortably, hoping that I would not be drawn too deep into discussion about the matter. The fact was that whenever snakes appeared Jack, as Curator of Reptiles, was the person *ex officio* who caught them. My achievements had been limited to once picking up a very small, timid, non-poisonous python in Africa.

There was a long pause.

'Well, good night,' said Clarence brightly, and he disappeared.

Charles and I settled down to our meal of tinned sardines followed by one of the pineapples which Clarence had brought us. Darkness fell and we climbed into our hammocks and prepared for sleep.

We were wakened by a loud 'Good night!' I looked up and found Clarence with the entire population of the village standing around the door of the hut.

'You tell these people what you tell me,' demanded Clarence.

We got up, repeated our stories, showed the light of the paraffin lamp in the camera viewfinder and played the recorder.

'We all sing now,' announced Clarence, organizing the rest of the villagers into an orderly group. Without the least trace of enthusiasm, they chanted a long dirge in which I thought I could detect the word 'hallelujah'. I remembered what Bill had told me.

'Why you sing Hallelujah chant, I thought this be Adventist village?'

'We all Adventists,' explained Clarence airily, 'and sometimes we all sing Adventist song, but when we *really* happy,' he added, leaning forward conspiratorially, 'we sing Hallelujah.' He brightened. 'Now we sing Adventist song because you ask for it.'

I recorded the chant and when it was finished, played it back to the villagers on a small speaker. They were entranced and Clarence insisted that now each member should perform a solo. Some sang guttural dirges and one man produced a flute, made from the shin-bone of a deer, on which he played a simple tune. This lengthy concert slightly embarrassed us, for we only had a limited number of tapes and our machine, being a small, lightweight, battery-operated model, did not carry an erasing device. If I recorded everything, I might waste all my valuable tapes on these relatively uninteresting party pieces and when I heard genuine spontaneous material there would be no tapes left. So I tried to record the minimum of each performance necessary to convince each singer that justice had been done to him.

After an hour and a half, the music came to an end and the villagers sat around the hut chattering in Akawaio, fingering our equipment and clothes and laughing among themselves. We could not join in the conversation and Clarence was involved in a heated discussion with another man outside the hut. We sat, ignored, wondering what was the polite thing to do and resigning ourselves to getting no sleep that night.

Clarence stuck his head in the door.

'Good night,' he said, beaming.

'Good night,' we replied, and all twenty of our guests without a word got to their feet and trooped out into the night.

The main occupation of the women was the making of cassava bread, thin flat cakes of which lay drying in the sun on the roofs of the houses and on special racks. They grew the cassava in plots between the village and the river. We filmed the women as they dug up the tall plants and removed the starchy tubers from among the roots. These they peeled and grated on a board studded with sharp fragments of stone. The juice of cassava contains a lethal poison – prussic acid – and to remove it a woman loaded the soggy, grated cassava into a matapee, a six-foot tube of extendable basketwork closed at one end with loops at top and bottom. When it was full she hung the matapee on a projecting beam of a hut. She passed a pole through the bottom loop and attached it to a rope tied to the hut post. Then she sat on the free end of the pole, and the laden matapee, squat and fat, stretched so that it became long and thin. In doing so, it squeezed the cassava and the poisonous juice trickled out at the bottom.

Squeezing out the poisonous juice from the grated cassava

The dry cassava was then sieved and baked. Some of the women used a flattened stone, but others employed a cast-iron circular plate, exactly the same as is used in Wales and Scotland in making girdle cakes. The flat disc of cassava bread having been cooked on both sides was put outside to dry.

We had been watching the process when Clarence came into the hut at a run.

'Quick, quick, Dayveed!' he yelled, waving his arms wildly. 'I find t'ing for you to catch.'

I ran after him to a log lying in a patch of low scrub close to his hut. By the side of it, I saw a small, black snake, about eighteen inches long, slowly swallowing a lizard.

'Quick, quick, you catch 'im,' Clarence cried enthusiastically.

'Well . . . er, I think we should film him first,' I said, procrastinating. 'Charles, come quickly.'

The snake, unconcerned by all the excitement, continued with its meal. The head and shoulders of the lizard had already disappeared and we could just see the tips of its front toes, pressed back against its body, projecting out of the corners of the snake's mouth. In girth, the snake was about a third of the size of the lizard, and to accommodate its enormous meal, the snake had unhinged its lower jaw. Even so, its little black eyes almost popped out of its head.

'Eh!' called Clarence to the world at large. 'Dayveed, he goin' to catch this bad snake!'

'Is this a *very* bad one?' I asked Clarence nervously.

'I don' know,' he replied with relish, 'but I t'ink 'e terrible bad.'

Meanwhile Charles was already filming. He looked up over his camera. 'I'd like to help,' he said smugly, 'but I must record this unique exhibition of gallantry.'

By now, the snake had reached the hind legs of the lizard. It was not eating it so much as crawling over and around its victim, for the lizard had remained in exactly the same position relative to the ground, while the snake slowly advanced towards its victim's tail. It did this by wrinkling its body into zigzags and then stretching straight in rather the same way as one threads a cord into pyjama trousers.

Most of the villagers had assembled in an expectant circle. The last tip of the lizard's tail disappeared and the little snake, grossly distended, began to crawl heavily away.

I had no excuse for further delay. Taking a forked stick I jabbed it down astride the snake's neck, so that the reptile was pinned to the ground.

'Quick, Charles,' I said, 'I can't do anything more unless you have a collecting bag.'

'Here's one,' Charles replied cheerfully, pulling a small cotton bag from his pocket. He held it open. With great distaste, I put a thumb and forefinger round the snake's neck, picked it up and dropped the wriggling creature into the bag. I heaved a sigh of relief and in as casual a manner as I could manage walked back to our hut.

Clarence and the spectators trotted behind.

'Sometimes maybe we find a big, big, bushmaster snake and then you show us how you catch *that* t'ing,' he chattered enthusiastically.

It was a week before I could show my capture to Jack.

'Non-poisonous,' he said tersely, handling it with complete unconcern. 'You won't mind if I let it go, will you?' he added. 'It's very common.' He put it on the ground beneath a bush. As I watched, it wriggled rapidly away through the undergrowth.

Late one evening, a young Akawaio boy trotted into the village. Over his shoulder he carried a blowpipe and in his hand he held a cloth bag.

'Dayveed – you want these t'ings?' he said shyly.

I opened the mouth of the bag, and peered cautiously inside. To my astonishment and delight I saw at the bottom, lying perfectly still, several tiny hummingbirds. I shut the bag quickly, and excitedly ran into our hut where we had a cage made from a wooden crate ready for any animal that might turn up. One by one I put the little birds inside. To our relief, they immediately took to flight, darted rapidly through the air, hovered and then jerkily reversed to settle on the thin perches with which the cage was fitted.

I turned to the boy, who had followed me.

'How you catch urn?' I asked.

'Blowpipe – and these t'ings,' he replied, handing me a dart. Its sharp point had been tipped with a little round pellet of bee's wax.

I turned again to the hummingbirds. The light blow from the blunted dart had obviously only stunned them temporarily and they were now busily flitting to and fro in the cage.

One of them was a particularly beautiful creature, not more than two inches long, which I recognized: before we left London, I had visited the Natural History Museum and had been captivated by one of the most delicate and gorgeous of all the hummingbird skins there. It was labelled *Lophornis ornatus*, the Tufted Coquette. What had been beautiful even as a stuffed bird was here a breathtaking spectacle of movement and exquisite colour. On top of its tiny head it flaunted a short crest of vertical topaz-red feathers. Beneath its needle-thin beak shone an iridescent emerald gorget and fanning out on either cheek was a sheaf of topaz feathers flecked with spots of emerald.

I was both entranced and dismayed, for although I had hoped to see this bird more than any other, we had decided that Jack should concentrate on collecting hummingbirds at Kamarang and we had not brought any of the necessary feeding equipment with us to the village.

Hummingbirds live mainly on nectar from forest flowers. In captivity they will readily accept a solution of honey and water enriched with milk extracts. As they only feed on the wing, special bottles with a cork at the top and a tiny spout at the bottom are needed to enable them to sip this substitute nectar. We had none of these things with us.

By now, it was dark and the little birds would not feed even if we had been able to offer them anything. We squatted in our hammocks and prepared a solution of sugar, hoping that it might provide them with enough sustenance. Laboriously we tried to improvise feeding bottles by boring holes in the base of a section of bamboo and inserting small spouts from the stem of another tree. The finished result seemed very crude and we went to bed rather despondently.

We were wakened in the middle of the night by a tremendous rainstorm. The roof of the church had great gaps in it and we leapt out of our hammocks to shift all our equipment and the hummingbirds in their cage to a dry spot. For the remainder of the night I slept only fitfully as the rain dripped down around me and collected in puddles on the floor. My single blanket became clammier and clammier. In my mind I could hear Bill saying that the seasons were deranged; that the rains might well begin early and that once they started they might continue for days on end without any real let up.

In the morning, with the rain still drumming on the roof and pattering on the floorboards of the hut, we did our best to persuade the hummingbirds to feed from our improvised bottles. We had no success; our substitute equipment was altogether too crude and the sugar solution rapidly dripped out of the bottle before the birds had taken any. We knew that they had to feed several times a day and that without regular supplies they would quickly wilt and die, like flowers without water.

With a wrench, we made the decision to release them, but having done so a weight lifted from our minds as the fragile things flew off through the door of our hut straight to the forest.

I sat in the doorway of the hut and brooded while Charles busied himself among the stores and equipment. Outside, I could see the village through squalls of rain, huddled forlorn and desolate under the dismal sky. If this was indeed the beginning of the rainy season, all our plans for filming in the Mazaruni Basin would have to be abandoned and all the trouble and expense of getting there would have been wasted. I thought miserably of how exultant Jack would have been to have seen the tufted coquette and the other hummingbirds we had just released, and of how foolish and shortsighted we had been not to bring any feeding bottles.

Charles joined me. 'I've made a few discoveries that might amuse you,' he said. 'First, that packet of sugar you've just emptied into your tea was our last. Second, I can't find the tin-opener. Third, the air is so damp that there's a great patch of fungus growing on one of the lenses of my camera, and fourth, I can't change that lens because the mounting has seized up.'

He looked pensively at the rain. 'If there's fungus growing on the glass of a lens,' he continued, 'it must be sprouting like mustard and cress on the exposed film. Not that that matters,' he added mournfully, 'because it's probably melted in the heat anyway.'

There was nothing for us to do but to wait until the rain stopped. I returned to my hammock. Dejectedly, I took out of my kit one of the few books we had brought with us – *The Golden Treasury*.

I read for a few minutes.

'Charles,' I said, 'have you ever felt that William Cowper, 1731–1800, had any particular message for you?'

Charles made a vulgar but dismal reply.

'You're wrong, listen,' I said.

> *O Solitude! where are the charms*
> *That sages have seen in thy face?*
> *Better dwell in the midst of alarms*
> *Than reign in this* 'orrible *place.*

5

Spirits in the Night

The rain continued intermittently for the last three days of our stay in Wailamepu. Although there were short periods of watery sunshine, serious photography was impossible, and we spent our time talking with Clarence, swimming in the tepid river, and watching the everyday life of the people. This was pleasant enough, but we were constantly nagged by the thought that precious time was passing and that there were still many interesting aspects of village life which we had not yet filmed.

On the seventh day of our stay, we packed up our gear in preparation for the return of the canoe. Clarence was helping us spread groundsheets over our pile of equipment to shield it from the rain dripping through the roof, when he straightened up and said conversationally, 'Kenneth arrives in half-hour.'

His confident statement mystified me and I asked him how he could be so sure.

'I hear engine,' he said, amazed that I should have asked. I put my head out of the hut door and listened. I could hear nothing but the swish of the rain on the forest.

Fifteen minutes later both Charles and I decided we could just distinguish the faint noise of an outboard motor, and in half an hour, exactly as Clarence had predicted, the canoe rounded the bend of the river with Kenneth at the tiller, bare-headed in the rain.

We left our friends at Wailamepu with regret, tempered by the pleasant anticipation of the dry clothes that awaited us at Kamarang. When we arrived there, we found that Jack's week had, on the whole, been more profitable than ours, for he had assembled quite a large miscellaneous collection of animals. There were numerous parrots, several snakes, a young otter and several dozen hummingbirds feeding very happily from glass bottles, the lack of which had forced us to release our tufted coquette.

We discussed plans for the week that now remained before our plane was due to return to Imbaimadai to collect us. It was decided that Jack should remain at Kamarang and that Charles and I should set out again on another canoe journey with the object of visiting as many villages as possible. We asked Bill's advice.

'Why not travel up the Kukui?' he suggested. 'That's fairly heavily populated, and most of the villages are unmissionized, so you might hear some Hallelujah chants. Take the smaller canoe, and when you return down the Kukui, carry on up the Mazaruni to Imbaimadai. We will go up in the big canoe with all the animals and meet you there.'

We set off the next day with the intention of spending our first night at Kukuiking, the village at the mouth of the Kukui. King George and another Amerindian named Abel came with us. The small canoe was heavily loaded with food, hammocks, a new supply of film, several empty cages ready for any animals we might find, and a large stock of blue and white glass beads with which to buy them. The colour of these beads was important, as Bill had told us when we bought them at his store. In the upper Kamarang, the inhabitants were very fond of red and pink as well as blue beads for the manufacture of their bead-aprons and other personal ornaments. On the Kukui they were more conservative, and blue and white beads were the only acceptable currency.

We reached Kukuiking in the late afternoon. Like Wailamepu it was a collection of simple wooden thatched huts in a clearing in the forest. The inhabitants stood morose and silent on the bank as we disembarked. As cheerfully as we could, we explained why we had come and asked if anyone had any pets which they

Abel in the bows of the canoe

would be willing to exchange for beads. One or two bedraggled little birds in filthy wicker baskets were reluctantly produced, and the villagers continued to regard us very suspiciously. This was unexpected after the genial and cheerful people we had known at Wailamepu.

'These people not happy?' I asked.

'The headman, he very ill,' King George replied. 'He lie in his hammock for many weeks now and the piaiman – the medicine man – he going to piai and cure him tonight. So they not so happy here.'

'How does he piai?' I asked.

'Well, in the middle of the night he call spirits down from the sky to come and make the headman better.'

'Will you ask the piaiman if he will speak with us?'

King George disappeared in the crowd and returned with a prosperous-looking man in his early thirties. Unlike the rest of the villagers, who were wearing either shabby European clothes or loincloths and blue bead aprons, the piaiman was dressed comparatively neatly in khaki shorts and a shirt.

He looked at us rather sulkily.

I explained that we had come to the village to make pictures and recordings to take back to our country and asked if we might visit his seance that night.

He grunted and nodded.

'Could we perhaps bring a small light to take photographs?' I asked. He looked up and said severely, 'Any man who show light when spirits in the hut – he die!'

I passed over the subject quickly and picked up my tape recorder. As I did so, I plugged in the microphone and switched it on.

'May I bring this thing, then?' I asked.

'What kind o' t'ing, that?' he said disparagingly.

'Listen,' I replied, and wound back the tape.

'What kind o' t'ing, that?' repeated the small loudspeaker rather tinnily. The suspicious look on the piaiman's face dissolved into a grin.

'You fine t'ing,' he replied, addressing the machine.

'You agree that I bring it tonight, so that it can learn the spirit songs?' I continued.

'Yes. I 'gree,' said the piaiman amicably and he turned on his heel and walked away.

The crowd dispersed and King George led us through the village to a small empty hut on the edge of the clearing. We dumped our kit and slung our hammocks. As the sun set, I practised loading and unloading the tape recorder with my eyes tightly shut. It was not as easy as I had imagined, and I was constantly getting lengths of tape entangled with the knobs and levers of the machine. Eventually I felt reasonably confident that I could change reels in total darkness, but as a safeguard I decided to go to the seance smoking a cigarette so that, initially at least, I should be able to solve any unforeseen difficulties by the light of its glow.

Late that night, Charles and I picked our way in the darkness through the silent village. The pointed silhouettes of the huts jutted black against the cloudy moonless sky. We entered the big hut to find it crammed with people. A small woodfire burned in the centre of the floor, illuminating the faces and bodies of the men and women who were squatting in it. In the semi-darkness beyond, we could just distinguish the dim white underbellies of occupied hammocks, one of which we knew contained the sick headman. King George was sitting on the wooden floor close to where we stood. Next to him we recognized the piaiman, squatting on his haunches and naked to the waist. In his hands he held two large sprigs of leaves, and by his side stood a small calabash full of what we later discovered to be salted tobacco juice.

We sat down close by him. I carried a lighted cigarette in my hand as I had planned, but the piaiman spotted it immediately. 'Is no good!' he said aggressively, so I meekly stubbed it out on the floor.

The piaiman gave an instruction in Akawaio; the fire was kicked out and someone hung a blanket over the door. The shadowy outlines of the people sitting around me disappeared into blackness. It was totally dark. I groped for the recorder in front of me and found the switch so that I should be ready to begin recording as soon as the seance began. I heard the piaiman clear his throat and gargle with the tobacco juice. Then the leaves began to rustle. The eerie noise grew louder and louder, like a drum roll, until at its loudest it resolved itself into a hypnotic rhythmic beating which filled the hut. The piaiman's voice rose above the noise of the leaves in a moaning chant.

King George, just behind me, whispered in my ear. 'Is calling a *karawari* spirit to come. He shaped like rope and all other spirits climb down him.' After ten minutes the invocation came to an end. There was silence, broken only by the heavy breathing of someone close to me.

A rustle sounded high in the roof and slowly descended, increasing in volume until it ended abruptly with a thump on the floor. A pause – a gargle – and then a quacking noise. A strained falsetto voice began singing. Presumably this was the *karawari*. The song continued for several minutes, when suddenly, the pitch darkness was stabbed by a spurt of flame from the dying embers of the fire. In its momentary light, I saw the piaiman, still close to me, his eyes shut and his face contorted, with beads of sweat lining his brow. The flame died almost immediately, but it had broken the tension and the chanting and rustling stopped abruptly. Two boys on my left chattered uneasily.

The leaf rustlings began again. 'The fire frighten the *karawari*,' muttered King George in explanation. 'He no come again. Piaiman now try to get *kasa-mara* spirit. He look like man and he bring a rope ladder.'

The chant continued and once again in the blackness we heard a rustle descending from the roof. Another gargle – and a loud announcement in Akawaio which was replied to in fairly tart terms by a little girl somewhere on our right.

'What do they say?' I asked King George into the darkness.

'*Kasa-mara* say he work hard,' he whispered, 'and that headman must pay well; and the girl, she say "He only pay if you make him better".'

The leaves were now thrashing wildly and seemed to travel nearer the headman's hammock. Soon the voice of several villagers began to join in with the spirit song, and someone beat time with thumps on the floor, until the song ceased and the rustles rose and faded away in the roof.

Another spirit arrived – more gargles – more songs. I felt almost suffocated by the stifling heat and the smell of sweating bodies in the pitch darkness of the hut. Every few minutes I had to change the tape on my recorder, but many of the spirit songs seemed repetitive and I did not record them all. After about an hour and a half, our initial awe began to wear thin. Charles, sitting by me, whispered in my ear, 'I wonder what would happen if you wound back the tape now and made the first spirit reappear!'

I was disinclined to experiment.

The seance continued for yet another hour – spirit after spirit descended from the roof, sang its song over the headman's hammock and departed. Most of them had sung in a ventriloquist's falsetto, but eventually a different spirit arrived chanting in a retching, gulping manner that was quite frightening to listen to. I heard King George's voice whispering 'This *bush dai-dai*. He very strong spirit of strangled man who came from topside in the mountains.'

The atmosphere became oppressively tense and charged with emotion. The piaiman, sitting a few feet away from me, was now quite feverish for, in the darkness, I could sense his position almost exactly from the heat of his body. The maniacal chant continued for several minutes and then abruptly stopped. There was a tense silence and I waited a little apprehensively in the darkness for what would happen next. The seance had obviously reached its climax. Was there perhaps going to be a sacrifice?

Suddenly a hot sweaty hand gripped my arm. I swung round, startled, but I could see nothing in the blackness. A man's hair brushed against my face. I was sure it was the piaiman and it flashed through my mind that the nearest white men were Bill and Jack, forty miles away.

The piaiman spoke hoarsely in my ear. 'All is finish. I go make water!'

The next morning we were visited by a deputation of villagers, led by the piaiman, now once again smart and smiling. He stepped up on to the bark floor of our hut, which was raised twelve inches above the ground, and sat down.

'I come to hear my spirits,' he said.

The villagers crowded into the hut after him and sat down

in a circle around the recording machine. There was not room for all who wished to hear this miracle, and an overflow audience stood in a semi-circle outside the door.

I connected the speaker to the recorder and began playing the tapes. The piaiman was delighted and as the music of the seance floated out into the sunlight it was greeted with gasps of approval, nudgings, hushings and occasional nervous giggles. As each spirit song ended, I stopped the machine and made notes as the piaiman told me the name of each spirit, its appearance and origin and its capabilities. Some were invested with fearsome powers; others were efficacious with only minor ailments. 'Man!' said the piaiman in an ecstasy of approval of one song, 'that very powerful – and good for curing a cough!'

In all, I had recorded nine of these spirit songs. The last few inches of the last tape ran out of the spool and I switched off the machine.

'Where the rest?' inquired the piaiman peevishly.

'I'm afraid this machine humbug bad in the dark,' I explained, 'and he no able learn all the songs.'

'But you no get the most powerful ones,' the piaiman said, looking petulantly at me. 'You no get the *awa-ui* or the *watabiara* and they fine spirits.'

I apologized again. The piaiman seemed slightly appeased.

'Would you wish to see the spirits?' he asked.

'Yes, very much,' I replied, 'but I thought that no man was allowed to see them and that they only came down into a hut during the night.'

The piaiman smiled confidentially.

'In day,' he said, 'they have different shape and I keep them buried in my hut. Wait. I fetch them.'

He returned holding in his hand a screwed-up piece of paper. He sat down on the bark floor once again and carefully opened the paper. Inside were a number of small polished pebbles. One by one he handed them to me, explaining the identity of each. One was a chip of quartz, another a long stick-like concretion, and another had four odd projections on it which he explained were the arms and legs of the spirit.

'I keep them in secret place in my hut, because these very powerful spirits and if another piaiman get them, then he can use them to kill me. This one,' he added gravely, 'is a very, very, bad one.'

He handed me a nondescript little pebble. I examined it with care and reverence, and passed it on to Charles. Somehow, between the two of us, we fumbled and the pebble dropped on to the floor and disappeared down a crack between the bark floorboards.

'He my most powerful spirit!' wailed the piaiman in anguish.

'Don't you worry, we'll find him,' I said hastily, scrambling to my feet. I threaded my way through the appalled spectators and, lying down, wriggled under the floor of the hut. The ground was very gravelly and, as far as I could see, was covered almost exclusively with pebbles exactly like the incarnate spirit.

Charles, kneeling down on the floor above me, stuck a twig through the crack down which the precious pebble had disappeared. I looked anxiously at the place beneath, but there seemed to be nothing to choose between any of the pebbles. I selected one at random and passed it up through the crack to Charles, who offered it to the piaiman.

'No good!' the piaiman ejaculated icily, throwing it to one side with disdain.

'Don't worry,' I bellowed from beneath the floor, 'I'll find him,' and passed out two other candidates. These met with the same treatment. During the next ten minutes, we proffered several dozen small pebbles. At last he accepted one and grunted grudgingly, 'This my spirit.'

I wriggled back into the sunlight, very dishevelled and covered in dust. The villagers seemed as relieved as we were that the spirit had finally been recovered, and as I sat there I wondered whether we really had returned the correct stone, or whether the piaiman had decided to accept an ordinary pebble lest the villagers should think that he had lost one of his most powerful weapons and therefore some of his prestige.

The piaiman carefully placed the pebble with the others in his screw of paper and walked back to his hut to re-bury them.

We left the village that afternoon to continue our journey up the Kukui. We never found out if the headman's health improved.

―――――

When we had first met King George several weeks earlier, we had been misled by his ferocious scowl into thinking that he was ill-tempered and surly, and he had not endeared himself to us by what we took as his irritating habit of demanding gifts. If Charles took out a packet of cigarettes, King George would hold out his hand and say peremptorily, 'Thank you for cigreet,' and then accept the gift not as a favour but as a right. This always led to a general distribution of cigarettes, which meant inevitably that we should be short by the end of our trip, for we had budgeted carefully and accurately in order that our loads of stores might be kept to a minimum. However, we realized after a few days that the Amerindians regarded most property as communal: if one man had something that his companions lacked, then it was only right that he should share it. If food were short, then we should split our tin of bully beef with everyone in the canoe and, if we wished it, the Amerindians, as their share of the bargain, would give us some of their cassava bread.

As we got to know King George better, we valued him as a charming and kindly companion. He was full of information about the river and knew it intimately. At first, however, we occasionally found it difficult to convey our exact meaning to one another, for though King George spoke a limited amount of pidgin English, his words did not necessarily mean the same thing to all of us. 'An hour' to King George was plainly only an indeterminate period of time, for if we asked him how long it would take to walk from the riverbank to a village in the 'back-dam', he nearly always replied, 'Eh, man! 'Bout one hour!' The unit of an hour was never divided or multiplied and 'one hour' turned out to be ten minutes on one occasion and two and a half hours on another. This, of course, was entirely our fault for asking the question 'How long?' for there was little reason why our units of time should mean anything to King George.

It was slightly more satisfactory to inquire 'How far?' The answer to this varied from "E no far' (which probably meant an hour's journey) to 'Man, 'e faaar, far 'way', which meant it would not be possible to reach the place that day. We soon learned, however, that the most accurate way of assessing distance was in 'points'. By a 'point', King George meant a bend in the river, but to translate 'nine points' into time required some knowledge of geography, for near its mouth the river was straight for distances of several miles, whereas on the upper head-waters it twisted sharply every few minutes.

King George was ever obliging and always endeavoured to do what we wanted, though his willingness sometimes had slightly unfortunate results.

'Do you think we could possibly reach that village tonight?' I once said to him, implying by my tone of voice that I hoped very much that we could.

'Well, man,' he replied, 'I t'ink we *mus*' meet it tonight,' and he smiled encouragingly.

We were still travelling up an uninhabited part of the river at sunset.

'King George,' I said severely, 'where dis village?'

'Eh! 'E faar, far 'way!'

'But you say we meet it tonight.'

'Well, man, we tried didn't we?' he said in an injured tone.

As we travelled up the Kukui, the river became littered with fallen trees. Some of them we were able to sail round as they lay only partway across the river; others were so long that they spanned the banks like a bridge, and these we were able to slide beneath. Sometimes, however, we would come to a giant tree lying almost submerged, which we could not avoid. Then King

George would drive the canoe at it with an open throttle and at the last moment cut the engine, swing the propeller column up out of the water in case it should foul, and so force the canoe halfway over the obstacle. We then had to climb out and, balancing on the slippery log, with the current tugging at our feet, haul the boat the rest of the way across.

We stopped at small settlements every few miles to ask for animals. There was no place we visited which did not have its complement of tame parrots hopping along the eaves of the huts or waddling irascibly around the village with their wings clasped behind their backs. The Amerindians, like us, value them for their bright colour and their ability to mimic human speech, and often, as we arrived, the birds would shriek abuse at us in Akawaio.

Adult parrots are difficult both to catch and to tame, so the Akawaio take the young chicks from their nests in the forest and rear them by hand. At one village, a woman gave us a nestling which she had only just obtained. It was a most appealing little chick with wide brown eyes, an absurdly large beak and a few scruffy feathers poking their quills through its otherwise naked skin. I could not bring myself to refuse it, but if I were to keep the charming creature I should have to take lessons on how to feed it. The woman laughingly told me what to do.

First, I chewed some cassava bread. As it saw me doing so, the little bird became tremendously excited, flapped its stumpy feather-less wings and jerked its head up and down in its enthusiasm for the coming meal. I then put my face close to it, whereupon, without hesitation, it stuck its open beak between my lips. It was now up to me to thrust the chewed cassava bread down its throat with my tongue.

This seemed a disgustingly unhygienic way of feeding any crea-ture, but the woman assured me that there was no other method of successfully rearing a parrot chick. Fortunately ours was quite old, and a week later it was able to eat soft banana by itself and relieve us of the responsibility of chewing cassava every three hours.

The parrot chick

By the time we were nearing Pipilipai, the village at the head of the river, we had bartered beads for macaws, tanagers, monkeys and tortoises as well as several unusual and brightly coloured parrots. The most unexpected of our purchases was a half-grown peccary, the wild pig of South America. The villagers who owned him seemed quite glad to pass him on to us for a comparatively small quantity of blue and white beads. At the time, it did not occur to us to wonder why. We soon found out.

We had not expected to acquire such a large creature as a peccary and we had no cages big enough for him, but he was quite tame and we naively decided to give him a little rope collar and attach him to a cross-stay in the bows of the canoe. This, however, was more difficult than it would seem, for the peccary, roughly speaking, tapered from his shoulders down to his snout and it was soon apparent that no normal collar would stay on him for one moment. We therefore tethered him by tying a rope harness round his shoulders and forelegs. This, we thought, would be enough to dissuade him from trampling over other things in the canoe. Houdini, as we very soon called him, did not share this view and no sooner were we on our way than he lifted his forelegs, one at a time, and with ease slipped out of the harness and picked his way down the canoe to begin eating the pineapples we had brought for our supper. We were disinclined to stop and make other arrangements to secure him, for we had to reach Pipilipai that night and our engine, as King George expressed it, was 'humbugging plenty', so for the next hour I did my best to restrain Houdini's explorations by clasping his bristly body in an affectionate embrace.

At last we reached Pipilipai. The village lay ten minutes' walk away from the river and was one of the most primitive settlements we had so far seen. All the men wore loincloths and the women bead aprons. Their few circular huts were ramshackle

A tame crested curassow

and carelessly built. Some lacked side walls, and all were built directly on the dry sandy ground instead of having floorboards like the huts at Kukuiking. Here, as at every other village, King George seemed to have a number of relatives and our welcome was cordial. There were parrots here too, but in addition, we saw a large crested curassow strutting among the huts. It was a glossy, black turkey-like bird with a handsome topknot of curly feathers and a bright yellow bill. We learned that he was destined for the cooking pot, but the villagers found our blue beads irresistible and gladly bartered him for six handfuls.

There were no vacant huts in the village so, with King George and Abel, we slung our hammocks in a hut that was already occupied by a family of ten. While Charles prepared the evening meal, I fondled Houdini and treacherously tied a new and elaborate harness round his shoulders as I did so. I then tethered him to a post in the centre of the village, put a pineapple and some cassava bread at his feet and exhorted him to lie down and go to sleep.

The night was not a good one. King George had not seen his relatives for some considerable time and long after nightfall he was chattering away, exchanging gossip. At about midnight a child suddenly began screaming and refused to be placated. Then one of the men climbed out of his hammock and re-stoked the fire in the centre of the hut. At last I managed to get to sleep, but it seemed that no sooner had I shut my eyes than I was being shaken by the shoulder and King George was saying in my ear, 'De hog! 'E loose.'

'We'll catch him when it's light,' I murmured, and turned over to go back to sleep. The child started howling again and the unmistakable stench of pig filtered up my nostrils. I opened my eyes and saw Houdini rubbing his back against a hut post. Obviously, no one would get any sleep until he was re-tethered, so I wearily swung my legs out of the hammock and called softly to Charles to come and help catch him.

For half an hour Houdini cantered in, out and round the hut while Charles and I, bare-footed and half naked, chased after him.

Finally, we collared him and re-tied him to his post. Houdini, apparently satisfied now that he had wakened the entire population of the village, gave a hollow chop with his jaws and settled down on the ground with a pineapple between his front legs. We returned to our hammocks to try and sleep through the last few hours that remained before daybreak.

The journey back downriver began well. We had constructed a large cage for the peccary from thin saplings bound together with strips of bark, and this was wedged in the bows of the boat. Houdini behaved perfectly for the first half-hour; the curassow, tethered by a piece of string round its ankle, perched peacefully on the tarpaulin covering our equipment; tortoises rambled about the bottom of the canoe, parrots and macaws screeched amicably in our ears, and the capuchin monkeys sat together in a large wooden cage, affectionately examining one another's fur. Charles and I lay back in the sun, staring into the blue, cloudless sky and watching the green branches of the trees slip past us.

But this did not last long, for soon we reached a difficult snag of logs. We climbed overboard and, with our heads down, began hauling the canoe over a submerged tree trunk. This was the moment for which Houdini had been waiting. Unknown to us, he had broken the fastenings of two of the lower bars of his cage, and in an instant he had jumped out of the canoe. I leaped into the water after him, nearly upsetting the boat, and after swimming a few yards, managed to catch him by the scruff of his neck. He kicked and splashed and squealed at the top of his voice, but at last I got him back into the remnants of his cage. Charles began the repair work while I stripped off my dripping clothes and laid them out on the tarpaulin to dry. Houdini, however, had obviously enjoyed his swim and was determined to have another, so for the rest of the journey one of us had to sit by his cage, re-tying the bars as quickly as he loosened them.

In the late evening we arrived at Jawala, King George's own village, half a mile up the river from Kukuiking. There we spent the night, having secured Houdini to a specially long tether, and quartered the rest of the animals in a derelict hut.

The next day was our last before we had to return to Imbaimadai. Most of the inhabitants of the village had been out hunting for the past week, but King George told us that they would return that day and sing Hallelujah in thanksgiving.

We had heard a great deal about this extraordinary religion which is peculiar to this part of South America and which, as its name suggests, is derived from Christianity. At the end of the last century, a Macusi from the savannahs visited a Christian mission. He returned to his tribe and then claimed to have visions during which he visited a great spirit called Papa, high in the sky. Papa had said that he required worship by praying and preaching and told the man to return to the Macusi people and spread the new religion which was to be called 'Hallelujah'. The new beliefs were also adopted by neighbouring tribes from the Macusi so that by the beginning of this century it had spread to the Patamona, Arecuna and to the Akawaio – all Carib-speaking tribes and very similar to each other. The missionaries apparently did not realize the Christian foundation of the religion. Generally, they condemned the beliefs that they found as being pagan and they wholeheartedly opposed them. No doubt their opposition was intensified when, as happened several times, new Hallelujah prophets declared that Papa had also predicted that white men would soon arrive preaching from books and offering contra-dictory versions of their own religion. To judge from the missionaries' fierce hostility, we thought it must retain many of the Amerindians' old pagan beliefs, and we wondered what to expect on the hunters' return – a slightly warped version of Christian worship or a barbaric ritual.

We asked King George if we might film the ceremony. He agreed and we settled down to wait.

After lunch we saw, in the distance, a woodskin canoe coming down the river. Thinking that it might be the first of the returning hunters, we strolled down to the landing to meet it.

The canoe moored, and we blinked in astonishment at the incredible figure who walked up the path towards us. From what we had heard, we had expected a slim lithe Amerindian in traditional clothes. Instead we saw an old man wearing a pair of brilliant blue linen shorts, a shrieking sports shirt spangled with aggressive multi-coloured designs representing Trinidadian steel-bands, and a Tyrolean felt hat complete with a white plume. This extraordinary apparition gave us a toothless grin, and stuck his hands in his ultramarine trousers.

'Man say you wish see Hallelujah dance. Before I dance, how much dollar you pay?'

Before I could say anything, King George, who was standing with us, began indignantly shouting a reply in Akawaio, gesticulating wildly with both arms. We had never seen King George so animated.

The old man took off his hat and twisted it nervously in his hand. King George advanced on him, still fulminating, while the old man retreated backwards to his boat. He climbed in hastily and paddled back down the river.

King George rejoined us, still panting, 'Man!' he cried with great sincerity, 'I told that *worthless* fellow that in this village we sing Hallelujah for the praise of God and that if he come to sing for money then that is not true Hallelujah and we don't want him at all.'

In the middle of the afternoon, the hunting party returned. Slung over their backs in woven baskets they carried loads of smoked fish, plucked carcasses of birds and kipper-brown joints of smoked tapir flesh. One man had a gun over his shoulder, and the rest were armed with blowpipes and bows and arrows. Quietly and without speaking to King George or anyone else in the village, they walked up to the main hut, the floor of which had been brushed and sprinkled with water in readiness for them. They carried their loads inside and stacked them around the centre pole. Still silent, they

76

left the hut and walked fifty yards along the path towards the river.
There, they formed up in a column three deep, and began chanting.
With slow rhythmic steps, two forward and one back, they advanced
in procession towards the hut. At the head of the column, three
young men led the singing and every few minutes turned to face
the rest of the dancers. Slowly they progressed up the path, lurching
forward and stamping to emphasize the simple rhythm of their
chant. As they entered the hut, the song and the rhythm changed
and they linked arms and circled the pile of fish and meat in the
centre. Occasionally, a woman from the village wandered into the
hut, and attached herself to the end of the procession. Several times
in the droning three-note chant I distinguished the words 'Hallelujah'
and 'Papa'. King George squatted on his heels, pensively fiddling
with a stick in the dust. The chant ended rather inconclusively and
the singers stood about looking abstractedly at the ceiling or examining the floor. Suddenly the men who had led the procession
began singing again and everyone re-formed into a line facing
inwards, each with his right hand on his neighbour's shoulder. After
ten minutes, the singers knelt down and, in unison, spoke a brief
and solemn prayer. They got to their feet and the man with the
gun walked over to King George, shook him by the hand and lit
a cigarette. The Hallelujah service was over and, strange though it
was, we were left with an impression of deep sincerity.

That night was to be our last in an Amerindian settlement.
I was unable to sleep. Towards midnight, I climbed out of my
hammock and walked slowly through the moonlit village. As I
approached the big round house, I heard the noise of voices and
saw the flicker of lights through chinks in the wooden walls. I
paused by the door, and I heard King George's voice say, 'If you
wish to enter, Dayveed, you very welcome.'

I stooped and walked inside. The hut was lit only by a large
fire which illuminated the smoked roof beams and the beautiful
curves of several dozen giant calabashes which were grouped on
the floor. Men and women lay in hammocks, crisscrossing from
beam to beam; others squatted on small wooden stools carved

Preparing to record Hallelujah

in the stylized form of a tortoise. Occasionally a woman, naked
except for her bead apron, rose and walked gracefully across the
hut, the firelight dappling her body. King George reclined in his
hammock, holding in his right hand a small mussel-like shell,
its halves tied together with a string passed through holes just
above the hinge. Reflectively he felt his chin, until he discovered
a bristle. Then he closed the rims of the shell firmly around the
hair and plucked it out.

The air was filled with a low conversation in Akawaio. One
man squatted by the enormous calabashes, stirring them with a
long stick and pouring out the pink lumpy fluid they contained
into a smaller calabash which was handed round to everyone in
the hut. This drink, I knew, was cassiri, and I had read of the
way in which it is supposed to be prepared. Its main constituent
is boiled grated cassava, but added to it is sweet potato and
cassava bread which has been assiduously chewed by the women

of the village. This addition of spittle is supposed to aid in the fermentation of the drink.

Soon, the small calabash was circulating among the people sitting close by me, and at length it was put in my hands. I felt it would be exceedingly impolite to refuse it, but at the same time I could not dismiss from my mind the method of its manufacture. I lifted it to my lips, and as I caught the acid smell of vomit that rose from it, my stomach heaved. I began drinking and realized that if I had to taste that initial sip again, I might well be unable to control my stomach, so with an effort, I held the calabash to my mouth until I had drained it. With relief, I handed back the empty bowl and smiled weakly.

King George leaned out of his hammock and grinned approvingly.

'Eh, you!' he called to the man in charge of the calabashes, 'Dayveed like cassiri and gets big thirst. Give 'im some more.'

I was immediately handed another brimming calabash. As quickly as possible, I poured it down my throat. On second acquaintance, I managed to discount the nauseating smell and decided that although cassiri was a bit gritty and lumpy, its actual bittersweet taste was not wholly unpleasant.

I sat listening to the conversation for another hour. It was a fascinating scene and I was tempted to run back to our own hut and fetch a flash camera. Somehow the thought was repugnant, it seemed an infringement of the hospitality which had been so generously offered to me by King George and his companions. Contentedly, I sat in the hut until the early morning.

6

Shanties on the Mazaruni

Georgetown seemed particularly attractive to us when we returned from the Mazaruni. We relished the thought of eating meals which were not emptied straight from tins and which we had not cooked ourselves, and of lying flat on a bed covered with clean white sheets instead of curling in a hammock under a damp crumpled blanket excavated from the bottom of a musty kitbag. But we also had a great deal of work to do: fresh supplies of food had to be bought and plans for the next trip to be made; the exposed film had to be sorted, repacked and sealed, and taken down to the city's cold store to be deposited in a refrigerated vault. The animals had to be transferred to the larger, permanent cages which Tim Vinall had had built in readiness for them, and some of them had to be taken to the Georgetown Zoo, which was already looking after the anteater and was now offering to take Houdini and the crested curassow as temporary lodgers.

Our next journey should have been to a remote area on the edge of the Amazon basin in the far south. There two missionaries were living and working with a very primitive and interesting Amerindian tribe. The only way for us to reach them, apart from a march through the forest which, there and back, would take six weeks, was to land in an amphibian plane at a point on a river some fifty miles away from the tribe, having previously arranged with the missionaries by radio that canoes and porters would be there to meet us. This was our plan, but to our dismay we discovered that the missionaries had been out of radio contact with Georgetown for the past three weeks. Their radio must have broken down, and so there was no way of warning them of our arrival. To land without advance preparations would be to maroon ourselves in uninhabited forest without guides, porters or any means of transport.

An alternative scheme, however, was already forming in our minds. We had been left a message by the manager of a mining company to say that the forest round one of his exploratory camps at Arakaka in the northern part of the country was particularly rich in animals and that at the camp itself there were several tame creatures which he would willingly give us.

We looked at the map. Arakaka lay at the head of the Barima River which ran roughly parallel to the northern boundary of Guyana and then swung north-west to empty into the estuary of the Orinoco. The map told us two other important facts. First, a small red symbol of an aeroplane, printed by the name 'Mount Everard' fifty miles downriver from Arakaka, showed us that we could reach that point at least by amphibian plane. Second, a cluster of red circles along the southern bank of the Barima indicated that there were many small gold workings; from this we inferred that there must be considerable traffic along the river and that there was therefore every chance of finding a boat which could take us up from Mount Everard to Arakaka.

We investigated further. The Airways told us that the only time during the next fortnight that the amphibian was free for charter was the following day, and the Docks told us that in twelve days' time a passenger ship would be returning to Georgetown from Morawhanna, a small settlement near the mouth of the Barima. If we were to go, we should have to go tomorrow. Unfortunately there was no means of warning the mining manager, for his only contact with his office in Georgetown was by radiophone, and while he could call his office, his office could not call him. We therefore left a message to be passed on the next time he called, saying that we would be arriving in Arakaka in three or four days' time. We booked passages for our return on the ship, SS *Tarpon*, and we chartered the amphibian.

The next day, we were in the air en route for Mount Everard, wondering whether these rather hurried and impromptu preparations would get us to Arakaka and back again to Georgetown within a reasonable time. After an hour's flying, the pilot yelled at us over his shoulder. 'That,' he screamed above the roar of the engine, 'is the best they can do for a "mount" hereabouts!' and he pointed below to a small hump rising about fifty feet above the forest of the flat coastal plain. Just beyond it flowed the Barima River, and at its foot clustered a few small houses, the first we had seen for seventy miles.

The pilot put the plane into a steep bank and shaped up for a landing on the river.

'I sure hope there's someone down there,' he bellowed, 'because if there ain't, there will be no one to moor the plane and no canoe to get you ashore, so we'll just have to take off again and go back the way we've come.'

'A fine time to tell us that!' muttered Charles.

The plane touched the surface of the river with a shudder, and through the spray spurting on the windows we saw to our relief a group of men standing on the jetty. At least we were going to be able to disembark. The pilot shut off the engines and shouted to the men to bring canoes. We unloaded all our gear and paddled over to the jetty. The plane took off with a roar, wished us luck with a tilt of its wings and disappeared.

The settlement at Mount Everard consisted only of six shacks grouped round a sawmill on the jetty. Hauled in the slipway close by lay a pile of huge, mud-blackened tree trunks that had been felled higher upriver and floated down to the mill. The jetty itself was covered in cones of fragrant salmon-coloured sawdust. The foreman of the mill, an East Indian, showed no sign of surprise at our unheralded descent from the skies, but simply and politely showed us to an empty hut where we could spend the night. We thanked him gratefully and asked him if there was any boat that might be going upriver towards Arakaka in the morning. He removed his peaked baseball cap and scratched his hair.

'No,' he said, 'I not think so. The only boat here is the *Berlin Grand.*' He pointed to a large single-masted wooden vessel lying with her sails furled by the jetty. 'She leave tomorrow with timber for Georgetown. But maybe something pass in two-three days.'

We settled down in our hut and prepared ourselves for a long stay. After supper, in the dusk, we walked down to the river. As we approached the *Berlin Grand*, we were hailed by the skipper, a hefty, elderly African man dressed in an oil-covered shirt and trousers, who was lying on the deck with his back against the mast. At his invitation, we went on board and met the other three members of the crew, all from the Caribbean, who were sitting with him enjoying the cool of the evening. We joined them and explained what we were doing on the Barima; they in turn told us about their life shipping sawn planks to Georgetown and bring back stores for the mill.

They talked not in Guayana pidgin, but in the rich dialect of the Caribbean, with that relish for the selection and precise placing of less common words that so enlivens Caribbean conversations. When we had returned from our Sierra Leone trip, I had given a great deal of drumming and chanting recordings to the section of BBC Radio's sound library that held traditional music from all round the world. I thought perhaps here was a chance to collect some Caribbean calypsos.

'Do you know many of the old sea songs?' I asked.

'Shanties? Yes man, I know plenty,' the skipper said. 'In fact my singin' name is Lord Lucifer. That's the man-devil. I have that name because when I get some of the right spirit inside me, I become a fiend to myself – a devilish man. And the first mate, he knows even more songs than I do, because he walk in the bush even longer than me. His name is Grand Smasher. You want to hear some shanties?'

I said that I would like to hear some very much and that I would also like to record them. Lord Lucifer and the Great Smasher held a muttered conference and then turned to me.

'OK, chief,' Lord Lucifer said. 'We sing. But you know, chief, I cain't remember the *good* songs, except there's a big supply of lubrication. You got dollar?'

I produced two. Lord Lucifer took them with a polite smile and called one of his crew.

'Present these,' he said solemnly, 'to Mister Kahn at the sawmill, with the compliments of the Berlin Grand and *insinuate*', his voice dropped to a whisper, 'that we require a supply of R-U-M.'

He gave me a wide toothless smile.

'With a little bit o' high spirit inside me, I'm a powerful singer.'

While the lubrication was being obtained, I set up the recorder. Five minutes later, the deckhand reappeared with sad news.

'Mister Kahn,' he said, 'gets no more rum.'

Lord Lucifer emitted a heavy sigh and rolled his eyes.

'We're goin' to have to work on a substitoot fuel,' he said. 'Request Mister Kahn to supply two dollars worth of Ruby Wine.'

When the messenger returned, he was clutching a vast number of bottles which he set down in ranks on the deck.

The Great Smasher picked one up and looked at it with distaste. In the centre of its garish label was a violently coloured design representing, rather inappropriately, a pile of lemons, oranges and pineapples. Above, printed in red capitals, were the words 'RUBY WINE' and beneath, more discreetly in small black letters, 'Port-type'.

'I's afraid we're going to need plenty of this stuff before we gits started really well on the good songs,' he said apologetically.

He pulled the cork, passed the bottle to Lord Lucifer, took one himself and, with a martyred air, manfully set about the task of lubrication.

Lord Lucifer wiped his mouth on the back of his hand and cleared his throat.

84

I'se shoutin' since de time I'm small,
I never like de t'ing dey call work at all.
Look me grandfather dead, going to work
Look me grandmother dead, coming from work
And me uncle, look he dead, working on a truck
So I don' see who the hell will get me to work.

We applauded.

'I know better ones than that, chief,' he said modestly, 'but I can't remember them just yet.'

He opened another bottle. Better ones soon began to arrive. The tunes of many of them I recognized as having been published in a collection of West Indian folk-songs. The printed words had seemed a little effete and lacking in coherent theme. Lord Lucifer's versions, however, differed considerably. Quite obviously, they were the originals from which the published ones had been derived, but they were so appallingly bawdy that as they rang out over the river, I was lost in admiration for the ingenuity of the folk-song collector who had managed to twist and trim the lyrics so that they became printable.

As the evening wore on and darkness fell, the crew and Lord Lucifer sang on. A chorus of frogs provided a honking accompaniment. The deckhand was despatched for further supplies of Ruby Wine. We learned what happened when 'Moskeeta married sandfly's daughter', and also of the no doubt apocryphal doings of Tiny McTurk's father in a shanty which began, 'Michael McTurk was a river navigator, and a great bush governor.'

The supply of Ruby Wine was dwindling but no further lubrication seemed to be necessary. Lord Lucifer and the Great Smasher were now singing in unison.

Madre, I'm tired of you, ah-ha,
Jus' because, you not really true, ah-ha,
For every time I walk down de strand,
I hear you in love, look, with some Yankee man.

We got to our feet and explained that we must go.

'Good night, chief,' said Lord Lucifer genially.

We walked rather unsteadily down the gangplank and up towards our hut while Lord Lucifer sang on.

Recording the shanty singers on the Berlin Grand

Next morning, the wharf was empty; the *Berlin Grand* had sailed at daybreak for Georgetown with a cargo of crabwood, mora and purple-heart timber. The settlement seemed deserted, the sawmill silent, sweltering in the moist oppressive heat. We walked up the little hill which gave Mount Everard its name with nets in our hands to see if we could find any animals. Little stirred in the broiling sun. A huge nest of leaf-cutting ants sprawled over the side of the hill, embracing its slope in a network of tracks; but no ants were to be seen. Occasionally, our attention was drawn by a rustle in the grass and we caught a brief glimpse of a lizard's tail. A few butterflies flew lazily and jerkily in front of us. Apart from these and the whirrs of crickets, there was no sign of life. If we were to be marooned at Mount Everard for long, it was clear that we should have to trek much farther into the forest away from the sawmill to find any animals.

Late in the afternoon, the brooding quietness was broken by the distant roar of an engine. Thinking it might be a launch, we ran down to the wharf to see if there was any chance of it taking us farther upriver towards Arakaka. The roar increased until round the bend at enormous speed came a tiny dugout canoe. It swept round in a wide flamboyant arc, casting a spectacular bow-wave. As it straightened out, the engine was cut off and the boat slid neatly up to the wharf. Two smart East Indian boys climbed out wearing singlets and shorts and white cloth peakless caps.

We introduced ourselves.

'Me, Ali,' said one, in reply. 'Him – Lal.'

'We wish to meet Arakaka,' Jack said. 'You able carry us?'

Ali, who was the spokesman of the pair, explained volubly that they were travelling upriver to cut wood, but that they were not going as far as Arakaka, that an extra load would not only slow them down but also increase the chance of sinking to danger point, and that anyway, they had not got nearly enough fuel to get to Arakaka and that even if they had, they would not have enough to get back. Clearly, it was impossible.

'But,' said Ali hastily, 'if you get plenty dollar, maybe we go.'

Jack shook his head and pointed out that the boat was exceedingly small, that it was uncovered and that we could not therefore protect our equipment from ruin by rain and that, now he came to think of it, we did not really want to reach Arakaka either.

Ali and Lal were delighted by this and we all sat down on the piles of sawdust on the wharf to savour to the full every move in this elaborate game of bargaining. At length, Ali agreed that, although he would undoubtedly lose a great deal of money on the deal, he would take us up to Arakaka the next morning for the paltry sum of twenty dollars.

That night there was a tremendous storm. The rain beat on the roof of our hut and cascaded through holes in the thatch on to the floor. Charles, having got up to make sure that the equipment was in a dry place, decided that as the noise of the storm would prevent him from sleeping anyway, he would seal everything in plastic bags in case a similar downpour caught us unprotected in the open canoe the next day.

In the morning, it seemed that we could not make the journey anyway, for Ali's canoe had filled with rain during the night and had sunk. It now lay on the bottom of the river with its engine under four feet of water.

Ali and Lal, however, were not at all put out, and had already begun salvage operations. With difficulty, they dragged the bows up on the bank. Lal began baling out water while Ali fished up the engine and hauled it ashore, water pouring from all its parts.

'Is all right,' he said. 'We get him to go soon.'

Nonchalantly, they began to dismantle it. Charles, who had a considerable knowledge of mechanical things, was doubtful. 'Don't you realize,' he said, 'that the coil is soaking wet? The engine will never start until it is completely dry.'

'Is all right,' Ali said again, unmoved. 'We cook 'um,' and taking off the dripping coil, he carried it over to a fire and put it on a bent plate of glowing metal. Then he removed the plugs from the engine and, together with other pieces of the mechanism, soaked them in petrol and set them alight. Every other removable piece of the engine was unscrewed and laid out on Lal's singlet to dry in the sun. This process seemed to have a horrid fascination for Charles, who sat watching and occasionally offering to help in what was obviously, to him, a completely new approach to mechanical repairs.

Within two hours, the engine was reassembled. With a flourish, Ali pulled the starting cord and to our astonishment the engine roared into life. Ali stopped it. 'We ready now,' he said.

Our misgivings over the size of the canoe were fully justified, for when we had loaded it with all our belongings and climbed in ourselves, there was barely an inch of freeboard, and the slightest movement by any one of us was sufficient to make the river water pour over the side. Our journey that day was therefore a little uncomfortable for we were very cramped and the enforced rigidity of our positions became extremely painful after a few hours. Nevertheless, we were very happy; we were on our way to Arakaka.

Even as we travelled, we saw more signs of animal life than we had ever done in the Mazaruni Basin. Morpho butterflies were very common and twice we saw snakes swimming in the river close by us; but we could do no more than slightly incline our heads to look at them for fear of capsizing the canoe. Occasionally we passed little clearings in the forest on the banks, with three or four half-naked Africans or East Indians standing watching us as we passed. Beneath them in the river lay logs which they had felled in the forest and tied together into rafts, ready to be floated down to the sawmill. Ali and Lal called out greetings and we slowly and noisily crawled past them. Once a small battered launch swept by us at speed and we spent an anxious few minutes baling hard to prevent our canoe from being totally swamped as we bobbed up and down in its wake.

Late in the afternoon we arrived at a small village. It looked pleasant and prosperous. Plots of cassava and pineapples had been laid out among the lush grass of the river bank and tall slender coconut palms grew between the solidly built huts. The Amerindians lined the banks watching us. Behind them and dwarfing them, stood two tall Africans.

We moored and climbed out; after five hours' travel, we were grateful for the chance to stretch our legs and move freely again.

Ali began unloading the canoe.

'This village Koriabo,' he said. 'Arakaka another five hours topside. We no take you farther. I t'ink canoe, 'e sink if we go again, and in this village, man gets launch. He take you Arakaka. Here – twenty dollar,' and to our surprise he produced the notes

and offered them to us. 'No,' said Jack, 'you bring us halfway
– you keep ten dollar.'

Ali flashed a smile. 'Thank you,' he said, 'now we go cut tree,'
and with Lal in the bows, he pushed the canoe out from the
bank. The tiny boat, freed of its enormous load, once more
surged speedily along the river and disappeared round the bend.

The taller of the two Africans walked up to us.

'My name is Brinsley McLeod,' he said. 'I get launch and take
you to Arakaka for ten dollar. It go down to Mount Everard to
get fuel this morning – maybe you see um – but it come back
tomorrow and then I take you.' We gladly accepted his offer and
walked up to the hut that had been allocated to us, well content
at the thought of travelling the next day in the powerful and
roomy boat that had swept by us earlier in the afternoon.

The next morning, as we were finishing breakfast, the other
African paid us a visit. He was considerably older than McLeod.
His face was scarred and deeply lined, and his eyes, the whites
bloodshot and yellowing, had a slightly wild look about them.

'Brinsley not say the truth,' he said darkly. 'That boat not come
back today, nor tomorrow, nor the nex' day. The men they stay
in Mount Everard drinkin' rum. Why you want to go Arakaka?'

We told him that we were collecting animals.

'Man,' he said gloomily, 'there's no cause to go Arakaka for
varmints like that. I get plenty o' them t'ings on my gold claim
in the bush here. Dere's camoodie, alligator, labaria snake,
antelope, numb fish. They are no use to me, you can take 'um,
'cause they humbug me bad. They *pests*.'

'Numb fish?' asked Jack. 'You mean electric eels?'

'Yes, plenty,' he said, heatedly. 'Small ones, big ones, some
o'them bigger than a canoe. They powerful bad varmints; they
can shock you through the boat, exceptin' you wear the rubber
long boots. One time they shock me and throw me on the
ground an' I lies there in the boat all dizzy in the head for three
days before I can get up. Yes, I get all them t'ings on my claim,
an' I take you there if you want to see 'um.'

We finished breakfast hastily and walked down with him to his canoe. As we paddled upriver, he told us more about himself. His name was Cetas Kingston and he had been prospecting for gold and diamonds in the forests of Guyana all his life. Sometimes he had made a strike but always the money had disappeared quickly afterwards, leaving him as poor as ever. A few years earlier he had discovered the claim he was taking us to now. This, he said, was the really good one, the one which would make him a rich man in a few years and enable him to give up working in the bush and settle down in comfort on the coast.

We turned off the main river into a side creek and soon we came to a post stuck in the swampy bank. A rectangular piece of tin, nailed on top, bore the crudely painted words 'Name of Claim HELL. Claimant C. Kingston', and beneath it a licence number and a date.

We climbed out of the canoe and followed Cetas along a narrow track into the bush. After ten minutes we stepped from the gloom of the forest into a sunlit area where the trees had been felled and the incomplete framework of a large hut had been built.

Cetas turned round to us, his eyes blazing. 'All round this place,' he said, sweeping his arm in a circle, 'there's gold in the ground. And it's none o' this no-good nugget gold. You find one of those t'ings one day an' then nothing for five year. No *sah*! In this ground, four feet down you come to the red gold-dirt, redder than blood. That's the true gold, an' all I got to do is to dig it out. Look, I show you.'

He seized the long-handled spade he had brought with him and frenziedly began to dig a small pit. Muttering to himself he hurled the spade into the ground. Sweat dripped off his tired face and soaked his shirt. At last he threw down the spade and groped in the bottom of the hole. He brought up a handful of rust-coloured gravel.

'There, you see,' he said hoarsely. 'Redder than blood.' He poked it with his forefinger and rambled on, almost ignoring us.

"Course, I am an old man, but I get two sons an' they fine boys. They ain't going to learn no trade. They'se going to come here and dig. And we'll plant cassava an' pineapple an' lime around the shack an' we'll bring in labour an' then we'll dig all this dirt up an' wash out all the gold.'

He stopped talking, threw the gravel in his hand back into the hole and got to his feet.

'I t'ink we go back Koriabo,' he said dejectedly, and walked down the trail to the canoe. He seemed to have forgotten that he had brought us to his claim to show us animals and to be obsessed with the sudden fear that though gold lay in the ground beneath his feet, he would not live to make the enormous fortune he had dreamed of all his life.

'Swim!' she said.

The children skipped down to the river. The capybara looked down their noses at us, turned and ambled after them. The children waited for the animals to arrive and then all four plunged in the water together and began to splash and wrestle, the children screaming with laughter.

Mama watched them with matronly complacency.

'I get all as babies together,' she said, and went on to explain that from the beginning of their lives the four infants had always bathed together and that now the capybara would not go into the river without the children.

The capybara playing in the river

We had told Mama that, like her, we were fond of tame animals and that we were hoping to take many back to our own country. Mama looked at the capybara. 'For me, them too big,' she said. 'You wish take them? I can get more.'

Jack was overjoyed at the offer, but a little uncertain as to how he would manage to transport such enormous creatures back to Georgetown. Eventually we arranged with Mama that we would try to get a cage built at Arakaka — if we ever got there — and collect the animals when we returned down the river.

Many of the other people in the village who possessed pets very understandably did not wish to part with them. One woman had a tame labba. It was a charming little creature with slender delicate legs like those of a miniature antelope. Like the capybara, it is a rodent and a relative of the guinea pig. Its coat was a rich brown, spotted with cream, and it gazed at us with lustrous black eyes as it lay in its owner's lap. The woman told us that three years earlier she had had a baby which had died in infancy. Soon afterwards her husband, hunting in the forest, had discovered a female labba with its young. He had shot the adult for food and brought the orphan back alive to his wife. She had taken the baby creature and suckled it at her own breast. Now it was fully grown. She stroked it affectionately. "E my baby,' she said simply.

That evening, we were surprised to hear the throbbing of an engine. As dusk fell a large launch came round the bend of the river and moored by the village. The East Indian captain in charge told us that he was bringing up stores and mail for the mining company and that the next day he would be continuing to Arakaka. He asked us if we would like to go with him, and we accepted readily: it seemed at last that we might reach our destination.

Early in the morning we carried our kit down to the launch. We explained to Mama that we would be back in four days' time to collect the capybara, and Brinsley promised to repair his boat so that he could take us down to Morawhanna when we returned. Most of the mining company's launch was occupied by freight, and there was one other passenger, a big cheerful African woman who was introduced to us as Gertie. Nevertheless there was plenty of room for us, and after the tiny dugout canoe and Brinsley's small boat, we thought it luxurious. We lay back in the bows and all three of us drifted off to sleep.

At four o'clock that afternoon we arrived at Arakaka. From the river it looked a charming and idyllic place; a string of small houses perched on the high bank, backed by tall sheafs of feathery bamboo swaying in the wind. When we landed, however, the charm dissolved. Two-thirds of the houses were stores combined with rum parlours and behind them, in muddy squalor, stood the dilapidated wooden shanties in which the villagers lived.

Fifty years ago, Arakaka had been a flourishing community of several hundred people. There had been rich gold mines in the bush nearby and it was said that the mining managers of those days used to drive with their wives in coaches along the main street. Now the gold mines were worked out and the street was grass. Most of the houses had fallen down, rotted and been reclaimed by the forest. An air of dissolution and degeneracy hung over the atrophied town as it mouldered in the heat. Near one of the shanties we found, submerged beneath a blanket of creepers, a weathered wooden table. Its feet were still embedded in decaying mortar and it stood on a platform of brickwork which was cracked and riven by the roots of the plants which concealed it. 'The hospital stood here,' we were told, 'and that's the old mortuary table.'

Although it was the middle of the afternoon the rum shops were already full and an old gramophone was blaring out tinny music. We went into one of the shops. A tall muscular young African man sat on a bench with an enamel mug full of rum in his hand.

'What you come up this way for, man?' he asked.

We said we were looking for animals.

'Well there's plenty here,' he said, 'an' I can catch 'em easy.'

'Splendid,' Jack replied. 'We will pay well for anything you bring us, but we are only here for a few days, so will you catch us something tomorrow?'

The man wagged a finger solemnly in Jack's face.

'Ah can't get anything tomorrow,' he said gravely, ''cause tomorrow I's going to be drunk.'

Gertie, our fellow passenger on the launch, strolled into the shop.

She leaned on the counter and looked hard in the Chinese storekeeper's eyes.

'Mister,' she said soulfully, 'the boys on the launch is telling me that there's plenty vampire bats up here. What can I do, 'cause I ain't got no mosquito net for my hammock?'

'You ain't bothered 'bout vampires, is you, ma?' said the African man with the enamel mug.

'I certainly am,' she replied stoutly. 'My psychological disposition is highly nervous.'

The man blinked hard. Gertie switched her attentions back to the storekeeper.

'Now, what you got to give me?' she said, with a simpering smile.

'I ain't got nothing to give, ma; but for two dollar I can sell you a lamp. That will keep the vampires away for sure.'

'Really, mister,' she said with exaggerated hauteur, 'I must add that my financial basis is very meagre.' She gave a laugh. 'Gimme a two-cent candle.'

Later that evening, my psychological disposition, like Gertie's, also became highly nervous, and for the same reason. We were staying in a decaying rest-house near the store. Jack and Charles went to sleep quickly underneath their mosquito nets, but I unfortunately had mislaid mine and for the past four days had been without one. Accordingly, because of Gertie's warning of vampires, I hung a lighted paraffin lamp at the end of my hammock. Ten minutes later, as I lay trying to sleep, a bat silently flapped in through the open window. It flew over my hammock, round the room, into the passage, back again under my hammock and out of the window. Every two minutes it came in and repeated this flight with unnerving regularity.

Without catching it, I could not be certain that it was a

vampire, but in such circumstances, zoological niceties are not necessary for conviction.

It did not seem to possess the elaborate leaf-shaped structure on its nose which many harmless bats have and which vampires lack, and although I could not see them, I felt sure it was armed with the pair of triangular razor-sharp front teeth with which vampires shave a thin section of skin from their victim. Having made the wound, they will squat by it and lap up the exuding blood. This they are able to do without disturbing a man's sleep, so that in the morning the only sign of their visitation is a blood-soaked blanket, though three weeks later the man may develop the dreadful disease of paralytic rabies.

A vampire bat

I found it difficult to believe implicitly in the storekeeper's assurance that vampires will never settle to feed where there is light, and my fears seemed to be justified when it suddenly settled in the far corner of the room and, in typical vampire fashion, began to scuttle around on the floor, its wings folded back along its forearms, so that it resembled some foul four-legged spider. I could stand it no longer. I reached below my hammock, picked up one of my boots and hurled it at the beast. It took to flight and disappeared through the window.

Within twenty minutes I was feeling almost grateful to the vampire, for the thought of it kept me awake for a long time and as a result I was able to achieve something which had become an obsession with me over the past few weeks: the recording of one of the most eerie sounds of the South American forest.

I had first heard this noise on our trip up the Kukui. We had pitched camp in the forest by the river and slung our hammocks between the trees. As we went to sleep the light of the stars twinkled through the leaves above. The ghostly shapes of bushes and creepers loomed around us. Suddenly, throbbing and echoing through the forest came an ululating yell rising in great crescendoes of blood-chilling loudness and then dying away to a moan like the sound of a gale wailing through telegraph wires. This terrifying noise was produced by nothing more fearsome than the howler monkey.

For weeks I had tried to record it. Every night that we were in the forest I had religiously fitted a microphone into a parabolic sound reflector, and loaded the recorder with new tape. Night after night we would hear nothing. Then one evening we would get to camp very late and very tired, and I would be too exhausted to set up the apparatus. That night, inevitably, I would be woken by the monkeys in full cry; I would jump from my hammock and frantically begin assembling the apparatus. As soon as everything was ready to be switched on, the chorus would stop. Once, on the Kukui, I thought that I had achieved success. The monkeys were so close that the noise was deafening and for once the recording apparatus was ready. I switched it on and for several minutes recorded the most

brilliant and terrifying howls I had heard. When the performance finished with two final yapping barks, I triumphantly
wound back the tape and roused Charles from his hammock
to hear it. The entire tape was blank; one of the valves had
broken during the day's journey.

Now, at last, thanks to the vampire bat, I was awake right at
the beginning of a chorus. The monkeys were probably half a
mile away but even so the noise was extremely loud. I lugged
the equipment out of the rest-house, set everything up and
carefully aimed my parabolic reflector in the direction from
which the sound was coming. After my previous experience I
did not play the tape back to Charles until the morning. We
listened to it together. The recording was perfect.

That morning the mining manager drove into Arakaka from his
camp twelve miles away in the forest. He had received our
message by radio but he was nonetheless a little surprised to see
us, and explained that he could not take us out to his camp that
day as his small truck would be piled high with the goods which
had been brought up on the launch. However, he suggested that
we might lunch with him the next day and promised to send
a truck in to collect us.

We spent the rest of the day rambling in the forest near the
town. Jack was hoping to find some interesting millipedes and
scorpions, and, coming across a low palm-like tree, he began
tearing off the dry brown leaf husks that wrapped round its
trunk. As he did so there was a loud hissing noise and a golden
brown furry creature, the size of a small dog, uncurled from the
upper part of the tree and hastily scrambled down the opposite
side of the trunk. It was on the ground and lumbering away
before we could get near it. It could not run fast, however, and
with a few strides Jack overtook it and picked it up by its stout,
almost naked tail. It hung upside down, glared furiously at us
with its little beady eyes, and hissed and dribbled through its

long curved snout. Jack was jubilant, for by sheer good fortune he had found a tamandua, the tree anteater.

We carried it back to the rest-house in triumph and while Jack began preparing a cage for it, we parked it in a tall tree close to the rest-house. The tamandua clasped the trunk with its forelegs and clambered up with rapidity and ease. When it was some twenty feet from the ground it stopped, turned round and gave us an angry look. Then it noticed that a few feet away hung a large globular ants' nest. Forgetting its irritation, it clambered towards the nest and, wrapping its prehensile tail round a branch just above, hung head downwards. With swift powerful swipes of its forelegs it ripped open the nest. A brown flood of ants flowed out of the gash and swarmed all over the tamandua, which, not in the least dismayed, stuck its tube-like snout into the hole and began licking up the ants with its long black tongue. After five minutes it absentmindedly began to scratch itself with its hindleg as it feasted. Soon it was scratching with one of its forelegs as well. Finally it decided that further food was not worth the penalty of additional ant stings, and it made a leisurely retreat. Its thick wiry fur was obviously not the complete protection against ants that it is often supposed to be, for at every other step the tamandua had to stop and scratch.

Charles and I were watching and filming its progress when it occurred to us that the task of climbing the tree to recapture the tamandua was not going to be a pleasant one. Angry ants were swarming all over the branches, and if the tamandua found their bites irritating, we should no doubt find them extremely painful. Fortunately the anteater solved the problem for us, for it scrambled down the tree and sat on the ground, rubbing its right ear with its back leg. The biting ants kept it so busy that it allowed Jack to pick it up and put it in the cage. There it squatted peaceably in a corner and set about removing ants from its left ear.

That night we went hunting with torches. In the darkness the forest seemed an eerie, mysterious place full of unseen yet

The tamandua

noisy activity. The texture of the sound varied from place to place; by the river, frogs filled the air with a metallic clinking, but as we moved further into the forest, the whirrs and chirps of insects became predominant. We quickly became accustomed to this unceasing chorus but the sudden crash of a falling tree or an echoing unidentifiable shriek brought my heart into my mouth.

Paradoxically, we were able to find things in the darkness which we would never have seen in daylight, for all animals' eyes act as reflectors, and as the beam of our torches fell on a creature looking in our direction, we saw two little lights shining back at us through the darkness. The size, colour and spacing of these eyes enabled us to make a guess at what animal we had found.

As we shone our torches over the surface of the river, we counted four pairs of glowing red coals – caiman lying almost submerged with their eyes just above the surface of the water. High in a tree we detected a monkey which, wakened by our steps, had turned to look at us. The reflection from its eyes disappeared momentarily as it blinked then vanished altogether and we heard a crash – it had turned its back on us and fled away through the branches.

Treading as silently as we could, we approached a thicket of bamboo, the stems creaking and groaning as they swayed in the darkness thirty feet above us. Jack shone his torch into the spiny tangle at the base of the clump.

'A good place for snakes,' he said with enthusiasm. 'You go round to the other side and see if you can scare anything out towards me.'

I picked my way with caution through the darkness and began beating against the bamboo with my machete. As I did so, the light from my torch fell upon a small hole in the ground.

'Jack,' I called softly. 'There's a small hole here.'

'I dare say there is,' he answered a trifle testily, 'but is there anything in it?'

Gingerly I knelt down and looked. From the depths of the hole three bright little eyes glowed at me.

'There most certainly is,' I called back. 'And what is more, it has got three eyes!'

Jack was by my side within a few seconds and together we peered down the hole. With the light of our two torches, we saw, crouching at the bottom, a black hairy spider as big as my hand. The eyes I had seen were only three of the eight that sparkled on the top of its ugly head. Menacingly, it raised its two front legs, exposing the iridescent blue pads with which they were tipped and giving us a clear view of its large curved poison fangs.

'A beauty,' Jack murmured. 'Don't let him jump out,' and he put his torch on the ground while he fumbled in his pocket for a cocoa tin. I picked up a twig and gently pushed it down the side of the hole. The spider lashed out with its front legs and pounced on it.

'Careful,' said Jack. 'If you rub any of that hair from its body, it won't live for very long.'

He gave me the tin. 'You hold that at the mouth of the hole and I'll see if I can persuade it to walk out.' Reaching over, he carefully pressed his knife into the ground so that the earth shook at the back of the hole. The spider revolved to face the new danger and retreated a few steps. Jack twisted the knife in the ground. The back of the hole crumbled and the spider suddenly ran backwards and landed straight in the tin. Quickly, I slipped on the lid.

Jack grinned with satisfaction and put the tin safely back into his pocket.

The next day would have to be our last at Arakaka, for our boat was due to leave Morawhanna at the mouth of the Barima in three days' time, and it would take us two days to get there. The mining company's jeep was due to come at midday to collect us and take us out to the camp twelve miles away. As we waited for it, we speculated with enthusiasm about the animals we should find when we arrived. But the jeep did not come on time and it was late afternoon before the manager drove in, full of apologies, to explain that the truck had broken down and had only just been repaired: it was now too late for us to visit the camp. We asked what animals might have been ours, had we been able to get there.

'Well,' he said, 'we did have a sloth but it died, and there was a monkey but that has now escaped. Still, I'm sure we could find a few parrots knocking about.'

We heard this with mixed feelings, disappointed at having come so far in pursuit of so little, but relieved that we had not missed anything spectacular by failing, at the last moment, to reach the camp.

The manager climbed back into his jeep and drove out of Arakaka. We were now faced with the problem of finding a boat to take us back down the river. We visited all the rum parlours in turn. There were many people who had canoes with outboard engines but everyone had a very good reason for not carrying us: the engine was broken, the canoe too small, there was no fuel, or the only man who really *understood* the engine was not in Arakaka at the moment. At last we discovered, sitting morosely in one of the rum parlours, an East Indian named Jacob. It would have been difficult not to notice him, for the tops of his ears sprouted long tufts of straight black hair which gave him the look of a dismal leprechaun. Jacob admitted that he had a boat, but said he could not take us. However, unlike everyone else he did not seem to be able to think of any good reason to support his decision and we pressed our case. We argued and haggled in the tobacco-laden atmosphere of the rum parlour while a gramophone screeched in our ears. At about half-past ten, Jacob's resistance finally broke down, and with deep gloom he agreed to take us to Koriabo in the morning.

We rose at six and were packed and ready to leave at seven. Jacob was nowhere to be seen. At nine o'clock he walked miserably up to the rest-house and announced that he had the boat and the engine was all ready, but that he had not yet been able to find any petrol.

Gertie, having nothing else to do, was standing nearby, following the conversation with interest. She looked at me sympathetically and heaved a heavy sigh.

'Man,' she said, 'ain't it terrible all this procrastination. It's real vexatious.'

By midday, however, the engine was fuelled and at last we set off downriver for Koriabo. The tamandua lay curled asleep in its cage with half an ants' nest lying by it for refreshment on the journey. Straddling the bows and overlapping an ample two feet on either side perched a large wooden cage which Jack had had made at Arakaka ready for the capybara.

It was extremely fortunate that we were travelling with the swift current of the swollen river, for Jacob's outboard engine was capricious to a degree and would brook no interference of any sort. If a small piece of floating wood temporarily blocked the inlet of the water-cooling system, or if we required it to run at anything but full throttle, then it stopped, and having stopped needed considerable coaxing to persuade it to start again. Jacob had only one method of doing this, and that was to pull the starting rope with maximum force as frequently as possible; the internal works of his engine were sacrosanct and must on no account be interfered with. His faith was always eventually justified, though on one occasion he had to pull the starting rope non-stop for an hour and a half. When it finally restarted Jacob, who had been gritting his teeth in subdued fury, betrayed no sign of triumph, but sat down at the tiller and relapsed into his normal state of deep gloom.

We reached Koriabo in the late afternoon and moored along-side Brinsley McLeod's launch. Jacob did not wish to stop the engine unnecessarily, so we unloaded as quickly as possible. Within ten minutes the canoe was empty, and Jacob, uncheered by the considerable achievement of completing the operation without stalling the engine, had dismally started on his way back to Arakaka.

We learned to our relief that Brinsley's launch was once more in running order, and although he himself was out of the village working his gold claim in the back-dam, we were assured that he would come back to the village at ten o'clock the next morning.

Somewhat to our surprise, he did. We enticed the capybara into their cage by filling it with overripe pineapples and cassava bread, and loaded it on board for our final day's travel down the Barima. The journey was a long one, for we stopped at all the little settlements we had visited on our way up to see if anyone had caught any animals, as we had asked. Several people had done so, and by the time we reached Mount Everard we had on board with us, in addition to the tamandua, the capybara and a snake, three macaws, five parrots, two parakeets, a capuchin

Enticing the capybara into a crate at Koriabo

monkey and, best of all, a pair of red-billed toucans. The bargaining inherent in the acquisition of all these creatures delayed us so much that darkness fell while we were still ten miles from Morawhanna and it was one o'clock in the morning when we finally nosed in alongside SS *Tarpon*, lying by the jetty at Morawhanna. We clambered up the gangway and went on board. After picking our way among the sleeping bodies that littered the decks, we eventually discovered the chief steward's cabin. He emerged, clad only in a pair of vividly striped pyjamas, but when he discovered that he had official functions to discharge, he jammed his peaked cap on his head and showed us to two cabins which, miraculously, had been reserved for us. We filled one with animals, and, at half-past two, the three of us wearily climbed into the bunks of the other.

When I next opened my eyes, it was midday; we were at sea and Georgetown lay on the horizon ahead.

8

Mr King and the Mermaid

Several surprises awaited us in the garage which held Tim Vinall's menagerie in Georgetown, for while we had been away on the Barima, our friends in other parts of the Colony had sent us more animals. The amphibian plane had recently visited Kamarang, and the pilot had brought back for us several parakeets and a tame red-capped woodpecker as a gift from Bill and Daphne Seggar. Tiny McTurk had sent a savannah fox and a bag containing several snakes. Tim, not content with the full-time job of looking after the collection, had also encouraged local people to bring in what animals they could find. Several had come from the Botanic Garden itself. A gardener had caught a pair of mongoose, families of which we had seen scampering over the lawns. Tim was glad to have them, although they are not strictly South American animals. They were imported many years ago from India by sugar planters in the hope that they would keep down the plagues of rats which caused so much damage among the sugar cane; since then they have increased in numbers so greatly that they now are one of the commonest animals along the coast. The gardens were also swarming with opossums, creatures which, like kangaroos, carry their newly born young in a pouch. I had been very eager to see this animal, one of the very few marsupials that are found outside Australasia, but I was sadly disappointed by them. Tim's two looked like enormous rats with pointed snouts almost naked of fur, long sharp teeth and repulsive scaly tails. They were without doubt the most hideous animals in the entire collection. Tim told us with glee that he had unhesitatingly christened them David and Charles.

Most remarkable of all these additions was a bad-tempered snivelling beast called Percy. Percy was a tree porcupine and like all members of the porcupine family, he was exceedingly ill-tempered. If anyone tried to touch him he screwed up his little face, rattled his short quills and hissed and stamped with rage, leaving it in no doubt that he would be delighted to use his long front teeth on anyone who came too near. His bristly tail was prehensile and with it he could grip on to branches as he climbed. Many tree-living animals are similarly equipped, but most of them – monkeys, pangolins, opossums and tamandua anteaters – possess tails which roll downwards. Percy's rolled upwards, a distinction he shared with, of all things, some mice living in Papua.

Percy, the tree porcupine

In spite of the addition of all these new creatures and the ones we had brought from the Barima, there still remained two important gaps in the collection, two extremely interesting Guyanese creatures which we had not yet caught. The first was a bird, the hoatzin. To scientists it is of particular interest for, unique among birds, it possesses claws on its wings. In the adult, these are useless and buried deep in the wing feathers, but they are fully functional in the unfledged chick, which uses its clawed wings as a second pair of legs to enable it to clamber in the branches round its nest. Fossils indicate that birds developed from reptilian ancestors. The hoatzin, with claws on its front limbs, is the only living bird to retain these characteristics, and the only place in the world where it is to be found is on the coasts of this part of South America.

The second animal we wanted so badly was a large seal-like mammal called the manatee, which spends its life in the creeks inoffensively browsing on weeds. Being a mammal, it suckles its young and does so by rearing out of the water, holding its single offspring to its breast, cradled in its flippers. It has been said that descriptions of the creature doing this, brought back by the first seamen to sail round the coasts of South America, gave rise to the legend of the mermaid.

We were told that both the hoatzin and the manatee were quite common in the Canje River a few miles down the coast from Georgetown. We had one week left in which to catch them, and so two days after we had returned from the Barima, we set off once more, this time by train, to the little town of New Amsterdam which lies at the mouth of the Canje.

Guyana became part of the British Empire only at the beginning of the nineteenth century. For several hundred years previously it had been governed by Holland, and as our train rattled along the coast signs of the Dutch occupation were still plain to see. The railway stations were named after the huge sugar estates they serviced – Beterverwagting, Weldaad, and Onverwagt. Many of the estates themselves owed their existence to the sea wall, far away on our left, which the Dutch built to convert the barren salt marshes into rich productive land, and in New Amsterdam itself, sweltering on the edge of the mile-wide estuary of the Berbice River, we saw, mingled with modern concrete buildings and wooden bungalows, a few elegant, white-painted houses – evidence of the dignity of the Dutch colonial architecture.

It seemed to us that the most likely people to help us in our search would be fishermen, so we went down to the harbour. Africans and East Indians sat mending nets and gossiping in their small wooden boats moored by the jetty. We asked if anyone could help us to catch a 'water-mamma' as the manatee is called locally. No one volunteered to do so, but everyone seemed to think that the man to help us was an African man named 'Mr King'.

He, it appeared, was a man of many parts. By profession he was a fisherman, but his strength was so prodigious that he was much in demand throughout New Amsterdam for all sorts of jobs such as pile-driving, which no other man could do properly. His recreation, we were told, was to 'wrestle' with cows. He was also a great hunter and knew more about the wild life of the district than anyone else. If there was one man who could catch a manatee, it was Mr King. We set off to look for him.

We found him at last sitting in the fish market, arguing with a trader about his catch. His appearance was startling enough to support his reputation. Immensely stout, he wore a brilliant red shirt, black pinstripe trousers, and on top of his mop of hair a very small black homburg hat. We asked him if he could catch a manatee for us. 'Well, man,' he said, fondling his luxuriant side-whiskers, 'there is plenty in these parts, but is very difficult to catch 'cause the water-mamma is the most passionate creature. When she get in the net, she fly into de most terrible passion an' throw herself about an' she so strong that she will bust open the strongest net.'

'What do you do about that?' Jack asked.

'There is only one t'ing to do,' said Mr King sombrely. 'When the water-mamma first get in the net, you must stroke the net ropes so that vibrations go down through the water to the water-mamma. If you do that right, she like it so much that she just lie there without moving an' go oooh!' Mr King emitted an expressive and ecstatic noise with a seraphic smile on his face. 'I only know one man who can do that,' he added, 'and that is me.'

We were so impressed by this display of expertise that we engaged Mr King on the spot. We had already arranged to hire a launch the next day, and Mr King agreed to bring two assistants and his net first thing in the morning to begin the hunt.

The launch was manned by an African captain and an East Indian engineer, and we soon discovered that neither of them held Mr King in as much awe as we had expected. After half an hour's travel up the Canje River, we saw an iguana high in one of the trees.

'There you are, Mr King,' said Rangur, the engineer. 'How'se about catching that?'

Mr King, with a lordly gesture, indicated that the launch should stop while he did so. He heaved his great weight into the dinghy, and with one of his assistants pushed off through the reeds to the base of the tree. The iguana, a splendid lizard about four feet long, lay immobile along a surprisingly thin branch, his green scales glinting in the sunlight fifteen feet above. M. King cut a tall bamboo and attached a noose to the end. This he hoisted and waved in front of the iguana's head.

'What do you do that for, Mr King?' called Fraser, the captain, with mock solemnity. 'You t'ink maybe he's going to climb down it into your hands?'

The reptile sat motionless, oblivious of what was happening beneath.

'Hey, Mr King,' taunted Rangur, 'that be a tame iguana that you tied to the tree last night to give yourself a good name.'

Mr King, however, was above answering such ignorant jibes

and instructed his assistant to climb the tree and guide the noose round the iguana's neck. The reptile responded to this by clambering lazily on to a higher branch.

'I think,' said Fraser, 'that the varmint is going to get the inclination to jump.'

Mr King exhorted his assistant to climb higher. For ten minutes the iguana, swaying in the breeze, permitted the noose to be dangled in front of it. Once it good-naturedly licked the rope as it swung close to its nose, but in spite of Mr King's encouraging shouts it refused to stick its head in the noose. Finally, the man in the tree got too close. The iguana's patience was exhausted, and with exaggerated unconcern it turned aside from the rope and dived gracefully through the air into the river. The last we saw of it was a muddy swirl in the depths of the reed thicket.

'I think,' said Fraser to the world in general, 'that he got the inclination.'

Mr King returned to the boat.

'There's plenty more of them t'ings,' he said. 'We are able to get plenty.'

The banks of the river were lined with high barricades of the giant mucka-mucka reed. Its stems, as thick as my arm and so spongily cellular that I could sever them with an idle swipe of my machete, rose straight and naked, until fifteen feet above the water they sprouted a few arrow-shaped leaves, and here and there a green fruit the size and shape of a pineapple. The leaves of the mucka-mucka we knew to be the favourite food of the hoatzins and we searched anxiously with the binoculars as we passed.

It was now midday. The sun beat down on us savagely. The metal fittings on the launch deck became too hot to touch. There was no breath of wind, and the leaves of the mucka-mucka hung motionless in the stifling heat, which shimmered over the river. Nothing stirred.

We saw our first hoatzin at about one o'clock. Jack's attention had been attracted by a muffled squawk rising from the mucka-

mucka. Fraser stopped the launch, and through Jack's binoculars we were just able to distinguish the outline of a large bird sitting panting in the shade of the reeds. As we drifted closer, we saw another and then a third. Soon we realized that the whole thicket was full of birds taking refuge from the broiling sun.

It was four o'clock before we had a clear view of a hoatzin. The sun had sunk considerably and the heat was less oppressive. We rounded a bend in the river and saw a party of six birds feeding on the mucka-mucka leaves. They were handsome, chestnut-brown creatures, the size of chickens, with heavy bodies and thin necks. Their heads were crowned with a tall spiky crest of feathers, and their glittering red eyes were surrounded by naked blue skin. As we approached, they ceased feeding and watched us, nervously jerking their tails up and down and uttering harsh grating cries. At last they flapped heavily a few feet farther into the thicket and subsided into its depth; but not before Charles had filmed them.

A hoatzin on its nest

We were thrilled to have seen these rare and beautiful birds, but our main interest was to watch the unique climbing behaviour of the chicks, so as we chugged slowly up the river, Jack continued to scour the reeds with his binoculars in search of a nest. We found one in the late afternoon. It was a flimsy platform of twigs hanging seven feet above the water in a thorn bush which grew among the mucka-mucka. Excitedly we clambered into the dinghy and paddled towards the bush. Two naked little chicks squatting on the nest peered over the edge, watching us. As we got closer their curiosity changed to fear and the scrawny wizened little creatures left the nest and groggily clambered up through the thorns, frantically gripping with their legs and their clawed wings. It was an astonishing and quite unbirdlike performance. As they clung to thin branches swaying above us, I stood up and gently reached towards them. Having demonstrated so perfectly their climbing ability, they then performed a trick which they alone among chicks can execute. They suddenly launched themselves into the air and dived neatly into the water nine feet below. They entered the water with hardly a splash and, as we watched, they swam energetically beneath the surface and disappeared deep into the thorny tangle.

We were disappointed that they had gone so soon for we had had no chance to photograph them, but we felt confident that as we had found chicks so easily on our first day there must be many more elsewhere on the river. Our search went on intensively for the rest of the time we spent on the Canje. We found several nests which contained eggs and one in particular was ideally suited to photography. As we approached it, the parent fluttered off heavily but soon returned, edging her way down the thin branch with her toes turned inwards. As she sat, she did not settle herself with a wriggle on to her eggs, but squatted uncomfortably on them in a seemingly haphazard way.

We returned to visit her many times over the next few days in the hope that her eggs might have hatched, but they had still not done so when the time came for us to return to Georgetown and we never saw any chicks other than those that had entranced us on our first day.

———

Early on the evening of our first day we reached a point on the river where it was joined by a small creek. This was the perfect place for catching manatee, Mr King told us. In half an hour the tide would turn and water would flow strongly down the creek into the Canje, carrying with it the lazy manatee which habitually grazed on its luxuriant weeds. All that had to be done to catch them was to stretch a net across the mouth of the creek. Accordingly, he plunged posts into the mud of the river bed on either side of the creek and stretched his net from one to the other. Then he sat in the dinghy alongside with his black homburg still in place, puffing at his pipe and awaiting the opportunity to display his remarkable rope-stroking skill.

After two hours he gave up. 'No good,' he said, 'the tide is not strong enough and no water come down. I know a better place where we catch 'em tonight.'

So the nets were hauled on board, the dinghy tied to the launch's stern and we moved on up the river. At dusk we reached the jetty of a sugar plantation. The overripe sickly-sweet smell of molasses drifted over the river as we moored. Rangur appeared from the galley with a steaming dish of rice and shrimps. After the meal Mr King explained with a martyred air that we could now go to sleep; he would catch the manatee during the night and show it to us in the morning. We, however, wanted to see how it was done, so we asked if he would rouse us before he began.

'Man,' he said, 'you don' want to come with me. I's going to be working at two-three o'clock.'

We assured him that we did, but it was only with the greatest reluctance that he finally agreed to wake us.

The insects on the river swarmed thicker than we had ever

seen them. There were sandflies, kaboura flies, mosquitoes and, a novel addition, large hornets. They swarmed in through the hatches and circled our lamp in a dense cloud. Others, not having found the entrance, collected on the outside of the portholes in such numbers that they completely covered the glass in an opaque scum. Charles, being in charge of our medicine chest, looked out a large pot of insect-repellent ointment ready for the night's operations. We hung our mosquito nets, climbed into our bunks and went to sleep.

It was Jack who roused us at two o'clock for the manatee hunt. We dressed carefully in long-sleeved shirts, tucked our trousers into our stockings and as an additional protection against the insects liberally daubed our hands and faces with ointment. We clambered aft to see if Mr King was ready. We found him lying on his back in his hammock, his mouth wide open, snoring stridently.

Jack shook him gently. Mr King opened his eyes.

'Wha did you do that for, man?' he said aggrievedly. 'Is the middle of the night. I is asleep.'

'What about catching the water-mamma?'

'Can't you see? Is too dark. There's no moon and I can't catch water-mamma in the dark, can I?' And with that he shut his eyes.

As we were up and dressed we decided that whether Mr King came or not we might as well do a little hunting on our own account. There were obviously plenty of caiman in the river for, flashing our torches over the inky surface of the water, we saw several pairs of telltale lights glowing back at us. We climbed into the dinghy, cast off and drifted silently down the river. Charles and I sat in the stern paddling as noiselessly as we could, while Jack squatted in the bows, torch in hand. Slowly we glided towards the reeds fringing the bank. There was nothing to be heard but the distant croaking of frogs and the occasional high-pitched whine of mosquitoes. Jack slowly moved his torch beam over the surface of the water. Then, abruptly, he ceased waving it and shone it steadily on a patch of reeds. He signalled to us to stop paddling.

Quietly we shipped our paddles and the boat drifted imperceptibly closer and closer to the reed bed. Soon we could distinguish in the torchlight the glistening scaly head of a caiman lying just above the surface of the water facing us. Holding his torch steadily in the beast's eyes, Jack slowly leaned over the bows. As he did so his foot touched a baling tin in the bottom of the boat. There was a faint clatter and a swirl in the water ahead of us. Jack sat back and turned to us.

'I think,' he said, 'that the varmint got the inclination.'

We began paddling again and within five minutes Jack had spotted another. Once again we glided towards it but when we were about ten yards away Jack switched off the beam of his torch.

'We'll forget about that one,' he said. 'Judging from the space between his eyes he is about seven feet long, and I'm not going to try to catch him barehanded.'

Soon, however, we had discovered a third. Once again we repeated the approach, sliding silently over the glassy black surface of the river, our attention riveted to the pool of light cast by Jack's torch and the two red unwinking lights in its centre.

'Come and hold my feet,' Jack whispered.

Charles moved quietly down the boat and clasped his ankles. As the boat drifted slowly towards the bedazzled caiman, Jack once again hung over the side. We got closer and closer until, from where I sat in the stern, the eyes of the caiman disappeared from my sight beneath the bows. Suddenly there was a splash, and a triumphant 'Got him' from Jack. He dropped his torch into the boat and hung over the gunwale, grappling with the caiman with both hands.

'Hang on for heaven's sake,' he called frantically to Charles, who was by now sitting on Jack's ankles and craning over the side himself. After tremendous splashings and gruntings Jack eventually leaned back into the boat, grinning. In his hands he grasped a snapping, struggling caiman over four feet long. While he held it by the scruff of the neck with his right hand, he tucked the creature's long scaly tail underneath his arm. The caiman hissed ferociously and opened its formidable jaws, exposing the leathery yellow inside of its mouth.

'I brought your kitbag along as I thought it might be useful,' Jack explained hastily to me. 'Would you mind passing it?' There seemed no time to argue, so I handed it to him. When I held it open, Jack carefully put the caiman inside and pulled the ropes of the bag tight.

'Well, that's something to show Mr King, anyway,' he said.

We spent three more days cruising on the Canje River with Mr King and his crew, searching for manatee. We set nets at night, we set them during the day; we set them in the rain and in sunshine, when the tide ebbed and when the tide flowed, but never did we see any sign whatsoever of our quarry in spite of assurances that each of these conditions in turn was essential for our purpose. Finally we had no more provisions to stay out longer, so dolefully we sailed back to New Amsterdam.

'Well, man,' said Mr King philosophically as we paid him off, 'I reckon we get bad luck.'

As we walked away along the jetty, an East Indian fisherman ran up to us.

'You the men that want a water-mamma?' he asked, ''cause I got one three days ago.'

'What did you do with her?' we asked excitedly.

'I put her in a small lake just outside the town. I can easily catch, if you want.'

'We most certainly do,' said Jack. 'Let's go and catch her now.'

The East Indian ran back along the jetty, loaded his net on to a hand-truck and collected three friends to assist him.

As our little procession wended its way through the crowded

streets, I heard the word 'water-mamma' being passed excitedly from person to person and by the time we reached the outskirts of the town and approached the meadow in which the lake lay, we had a large shouting crowd trailing behind us.

The lake was wide and muddy, but fortunately it was not deep. Everyone squatted on the banks and silently stared at the water, searching for a sign of the mermaid's position. Suddenly someone pointed to a mysteriously moving lotus leaf. It crumpled and vanished beneath the surface, and a few seconds later a brown muzzle appeared above the water, emitted a blast of air from two large circular nostrils and disappeared.

"He there. He there,' everyone shouted.

Narian, the fisherman, marshalled his forces. With his three assistants, he jumped into the water. Holding the long net stretched between them, he arranged them in a long line across the small bay where the manatee had been seen. Slowly, chest deep in the water, they advanced towards the bank. As they approached, the manatee betrayed her position by once more coming up for air. Narian yelled to the men on the two ends of the net to wade quickly to the bank and climb out so that the net formed not a straight line, but an arc. Now thoroughly disturbed, the manatee rose closer to the surface and rolled over, giving us a view of her great dun-brown flank.

A gasp of astonishment and pleasure rose from the crowd. 'She big thing! Man, she *monstrous!*'

Excitement gripped Narian's assistants on the bank and, enthusiastically aided by some of the onlookers, they began feverishly to haul in the net hand over hand. Narian, still wading in the lake, shouted furiously above the hubbub.

'Stop pullin',' he yelled. 'Not so fast.'

No one took the slightest notice.

'Hundred dollar, the net,' Narian screamed. 'He go bust if you not stop pulling.'

But the crowd, having seen the manatee's flank once again, were obsessed by the desire to land her as quickly as possible, and they continued to haul in the net until the manatee lay enmeshed in the water just below the bank. She was obviously a very big one, but there was no time to see more for she suddenly arched her body and thrashed with her enormous tail, soaking everyone in muddy water. The net broke and she disappeared. Narian's fury exceeded all bounds, and he scrambled on to the bank and wrathfully demanded payment from everyone standing nearby for repairs to his net. In the ensuing clamour, it did not seem appropriate to suggest that since our mermaid appeared to be a particularly passionate one, we should go and find Mr King so that he could stroke the ropes and pacify her the next time she was netted. The argument proceeded and everyone seemed to forget the manatee except Jack, who wandered off along the bank, tracing her course by swirls in the water.

At last the noise subsided. Jack called to Narian and pointed to where he had last seen the manatee.

Narian walked over, grumbling loudly, with a long rope in his hand.

'Those mad men,' he said contemptuously. 'They bust my net and he worth a hundred dollar. *This* time I going in the water an' tie a rope round her tail so she *can't* escape.'

He jumped into the lake again and waded to and fro, feeling for the manatee with his feet. At last he found her lying sluggishly on the bottom, and with the rope in his hands he bent down until his chin was just above the surface. He remained in this position for a few minutes as he groped in the water. Then he straightened and began to say something when the rope whipped tight in his hands, and pulled him flat on his face. He struggled to his feet, spat out the muddy water and happily brandished the end of the rope.

'I still got her,' he called.

The manatee, having passively allowed the rope to be tied round her tail, now realized her danger, and she reared to the surface, splashed and tried to bolt. This time Narian was ready for her and skilfully he managed to lead her towards the bank.

His chastened assistants once more encircled the manatee with the net and Narian scrambled up on to the bank with the rope still in his hand. The men on the net pulled, Narian heaved, and slowly, tail first, the mermaid was hauled ashore.

On land she was not a pretty sight. Her head was little more than a blunt stump, garnished with an extensive but sparse moustache on her huge blubbery upper-lip. Her minute eyes were buried deep in the flesh of her cheek and would have been almost undetectable if they had not been suppurating slightly. Apart from her prominent nostrils, therefore, she possessed no feature which could give her any facial expression whatsoever. From her nose to the end of her great spatulate tail she was just over seven feet long. She had two paddle-shaped front flippers, but no rear limbs, and where she kept her bones was a mystery for, robbed of the support of the water, her great body slumped like a sack of wet sand.

Narian with the manatee

She seemed entirely indifferent to our exploratory prods, and allowed herself to be turned over without so much as a wriggle of protest. As she lay motionless on her back, her flippers fallen outwards, I became worried that she had been injured during her capture, and asked Narian if she was all right. He laughed. 'This t'ing *can't* die,' he said, and splashed some water on her, whereupon she arched her body, slapped her tail on the ground and then returned to immobility.

The problem of getting her back to Georgetown was solved for us by the Town Council of New Amsterdam, who lent us the municipal water-lorry. We tied rope slings round her tail and beneath her flippers. Narian and his three assistants hoisted her from the ground and staggered across the meadow to where the lorry was parked.

Sagging between the slings, her flippers hanging down limply, and dribbling slightly from beneath her vast moustache, she seemed comfortable but hardly looked alluring. 'If any sailor ever mistook *her* for a mermaid,' Charles said, 'I reckon he must have been at sea for a very long time.'

9

Return

O ur expedition had come to an end. Jack and Tim were to bring the animals back to London by sea, but Charles and I had to return immediately by air to begin work on the film. Before we left Jack gave us a large square parcel. 'Inside this,' he said, 'there are a few nice spiders, scorpions and one or two snakes. They are all in sealed tins with tiny air holes so there's no possibility of them escaping, but try and keep them with you in the cabin so that they don't catch cold. And will you also take this young coatimundi kitten?' he added, passing me a delightful furry creature with bright brown eyes, a long ringed tail, and a pointed inquisitive snout. 'He's still on a milk diet, so you will have to feed him from the bottle every three or four hours on your way back.'

Charles and I climbed into the plane with the parcel and the coatimundi in a little travelling basket. The kitten was the object of a great deal of interest. As we flew over the islands of the Caribbean, a lady came to fondle him. She asked what sort of an animal he was, and how we came by him, and gradually we had to explain that we had been on an animal-collecting expedition. She looked at the box by my feet.

'I suppose,' she said with a smile, 'that that is full of snakes and other creepy-crawlies.'

'As a matter of fact,' I said in sepulchral tones, '*it is*,' and we all laughed uproariously at such an absurd suggestion.

The coatimundi behaved very well for the first part of the journey, but as we began flying north towards Europe he refused his milk. Fearing that he might catch cold, I tucked him inside my shirt, where he nuzzled beneath my arm and slept peacefully. I tried to persuade him to feed again in Lisbon, and once more at Zürich, but though we heated the milk and even tempted him with mashed bananas and cream in a saucer, he still declined to feed. We arrived in Amsterdam at one o'clock in the morning. The London plane left at six. Charles and I settled down to wait on the long leather couches of the airport foyer. Our little kitten had not fed now for thirty-six hours and we were becoming very anxious about him. We searched our memories trying to recall what is the favourite food of a coatimundi, but we could only remember that they were described in the natural history books as being 'omnivorous'.

Charles had a brainwave. 'What about some worms?' he said, 'He might be tempted if they were nice and wriggly.' I agreed, but neither of us was clear as to where we could get any worms at four o'clock in the morning in Amsterdam. Then it occurred to us that the Dutch, proud of their flowers, had surrounded the airfield with beautiful beds of plants which were now in full bloom. Leaving the kitten with Charles, I walked out on to the airfield, and in the glare of the floodlights I surreptitiously waded into the flowerbeds. Airport officials walked within a few feet of me as I dug in the soft earth with my fingers, but no one took the slightest notice and after five minutes I had over a dozen pink, wriggling worms. I took them back in triumph and to our delight the little coatimundi ate them greedily. When he had finished, he licked his lips and plainly asked for more. We made four more trips to the tulip bed before he was satisfied. Six hours later we handed him over, kicking lustily, to the London Zoo.

Meanwhile back in Georgetown a great amount of work still remained to be done to get the animals ready for the long voyage home. The last few weeks of our trip had been clouded by Jack's increasing ill health. Slowly it became apparent that he had contracted an extremely serious paralysing illness, and a few days

Coatimundi kittens

after we left him doctors in Georgetown recommended that he should be flown home as soon as possible to see a specialist in London. John Yelland, the Curator of Birds in the Zoo, flew out to Georgetown to take Jack's place and help Tim Vinall bring the collection to London by sea.

This was an arduous and complicated task: to ensure that the manatee had a comfortable trip, they arranged for a special canvas swimming bath to be erected on one of the decks of the ship; to cater for the enormous appetites of the animals they took on board a stock of provisions which included 3,000 lbs of lettuces, 100 lbs of cabbages, 400 lbs of bananas, 160 lbs of green grass and 48 pineapples; and to keep the collection clean and well fed on the nineteen-day voyage, Tim and John had to work unceasingly from dawn to dusk.

It was some weeks before I was able to go to the Zoo to see the animals again. I found the manatee swimming lazily to and fro in a crystal clear pool that had been specially built for her in the Aquarium. She was now so tame that when I leaned over and dabbled a cabbage leaf in the water, she swam to the side and took it from my hand. The little parrot that we had been given on the Kukui was now fully fledged and almost unrecognizable, but I convinced myself that he knew me, for when I talked to him he jerked his head up and down exactly as he had done when I had been feeding him chewed cassava bread from my mouth months before. The hummingbirds looked magnificent, darting and hovering among tropical plants in a specially heated house. Percy, the porcupine, I discovered curled up asleep in the angle of a branch, still with his unmistakable sour expression on his face.

When I found the capybara they were just about to leave for a large paddock in Whipsnade, the Zoo's country estate; they whistled and giggled and sucked my fingers as enthusiastically as they had done on the Barima. The anteaters still flourished on their diet of raw minced meat and milk, and in the Insect House I discovered that the spider we had caught at Arakaka had given birth a few days after it arrived to several hundred tiny young which were now fast growing up.

It took me some time to find Houdini, the animal that had caused me personally more trouble than any other. When I at last discovered him he had his head down noisily champing and guzzling in a large dish of swill. I leaned over the wall of his paddock and called him several times. He ignored me completely.

BOOK TWO

Zoo Quest for a Dragon

10

To Indonesia

The organization of an enterprise, eventually to be dignified by the title of expedition, should by rights imply many months of detailed planning. There should be schedules and permits; lists, visas and itineraries; enormous piles of carefully labelled baggage and equipment, and a chain of transport, the first link of which is a great cargo vessel and the last a queue of bare-footed porters. But we had made no such preparations for our trip to Indonesia and when Charles Lagus and I, with tickets for Jakarta in our pockets, climbed on board our plane in London, I confess I rather wished we had.

Neither of us had been in the Far East before, neither of us spoke Malay, and we had no contacts with anyone in Indonesia. Moreover we had decided, weeks earlier, not to take vast loads of stores with us on the grounds that two people can nearly always find enough to eat in places where an expedition of ten would starve. For similar reasons, we had not made any advance arrangements as to where or how we should sleep during the next four months. We had, it is true, visited the Indonesian Embassy in London. The officials there had received us most courteously and had promised to send letters to authorities in Indonesia, telling them of our expedition and asking for help. When we revisited the Embassy on the eve of our departure, however, we discovered that owing to a misunderstanding about dates, the letters had not yet been sent. As one of the officials put it, the easiest course in the circumstances seemed to be to take the letters with us and post them when we got to Indonesia.

Indonesia straddles the equator, stretching three thousand miles from Sumatra in the west to the western half of New Guinea in the east. It embraces Java, Bali, Sulawesi, the major portion of Borneo and many hundreds of smaller islands sprinkled between. From east to west it is as broad as the United States of America. Our plan was to travel widely throughout these islands, filming not only animals but also the people and their everyday life. Our ultimate objective was to reach a small islet, twenty-two miles long and twelve miles wide, lying almost in the centre of the archipelago, named Komodo. We were going there to seek one of the most remarkable creatures alive today, the largest lizard in the world.

For many years before the existence of this monster was scientifically confirmed, there had been rumours of an awesome dragon-like creature living on Komodo. It was said that it had enormous claws, fearsome teeth, a heavily armoured body and a fiery yellow tongue. These stories had been brought back by native fishermen and pearl divers who were the only people to sail among the dangerous reefs which surrounded the then uninhabited island and which made it one of the least accessible islands in the whole area. In 1910, an officer of the Dutch Colonial Infantry took an expedition to Komodo. He found that the stories were correct and to prove them he shot two of the giant lizards, brought their skins back to Java and presented them to a Dutch zoologist named Ouwens. It was Ouwens who first published a description of this astonishing creature and named it *Varanus Komodoensis*. The world at large promptly christened it the Komodo dragon.

Subsequent expeditions discovered that the creature was carnivorous, living on the flesh of the wild pig and deer which abounded on the island. It undoubtedly fed on rotting carrion, but it also seemed likely that it actively hunted its prey, killing it with a swing of its huge muscular tail. Specimens were found not only on Komodo but also on the neighbouring islet of Rintja and on the western tip of Flores, a nearby island. But the dragon occurs nowhere else in the world. This restricted distribution is a puzzle. The giant lizards are almost certainly descendants of the even larger prehistoric lizards whose fossil remains have been found in Australia. The most ancient of these are estimated to be about sixty million years old. Yet strangely, Komodo is a volcanic island of comparatively recent origin. Why the dragons should exist only here and how they reached the island are unanswered problems. As Charles and I sat in our plane, flying eastwards, the question of how we ourselves should reach Komodo seemed almost equally unanswerable. No one had been able to tell us in London. We hoped we should find an answer in the capital of Indonesia, the Javanese city of Jakarta.

The buildings of Jakarta are not in the least Oriental. The rows of neat white pantiled bungalows, the ferro-concrete hotels, the garish cinemas looking like monstrous petrified balloons, the occasional older building with an austere classical portico which survives from the Dutch colonial era – all these closely resemble modern tropical cities in others parts of the world. The people of Jakarta, however, did not seem as extensively Westernized as their city's architecture. Many of the men were dressed in sarongs, simple ankle-length skirts of local cloth; most wore a *pitji*, the black velveteen forage cap which, though originally part of the Muslim dress, had been consciously adopted and sponsored by the young republic as a symbol of national unity and was worn by Indonesian peoples of all races and religions.

The majority of the crowds in the streets were undoubtedly poor. Pedlars trotted along the gutters carrying on their shoulders long flexible poles with enormous loads suspended from each end – bales of cloth, racks of crockery and often a glowing brazier which its owner would set down at a moment's notice to cook a customer a quick meal of *saté*, delicately seasoned morsels of meat skewered on a splinter of bamboo. Standing in ranks at the roadside and dodging hazardously between the raucous American cars and the clanging trams were *betjaks* – the tricycle version of the rickshaw. Each of them was decorated with paintings of lurid landscapes and fearsome monsters, and carried beneath its seat a length of elastic, stretched between two nails, through which the wind hummed loudly and cheerfully as the *betjak* was pedalled at speed. Many of the main streets ran alongside canals, which the Dutch seemed compelled to build in every land they colonized. Lines of women sat on the banks washing fruit or clothes or themselves, some swimming in the water and others unabashedly using the canal as a lavatory.

Jakarta, in short, was noisy, crowded, bustling, in places squalid, and everywhere very, very hot. We longed for the moment when we could leave it.

———

We realized, however, that we should have to spend several days in the town paying courtesy calls on Government offices and obtaining formal sanctions for our plans, but we felt that, in view of the letters from the London Embassy, our way would be comparatively smooth. Not even in our moments of deepest pessimism did we imagine that we should have to spend more than a week in Jakarta. In retrospect, I realize that the difficulties we met were only to be expected. The newly created republic was facing incipient revolutions throughout its territories and, indeed, less than nine months later these flared up into open insurrection. We were foreigners, markedly similar to the Dutch who only six years earlier had been expelled as a colonial power from the country after months of savage fighting which produced a great deal of brutality on both sides. We were asking for permits to take film cameras and recording machines to remote areas of the republic which few of the Jakarta officials had even heard of. Furthermore – and this was probably our worst crime – we were in a hurry. Day after day in the sweltering heat we trailed round Government offices. To clear our equipment through the Customs, one of us had to report to the bonded warehouse every morning for a week. We were told that we must have financial clearances, military permits, police passes, letters from the Ministry of Agriculture and Forestry; we must have our plans approved by the Ministry of Information, the Ministry of Home Affairs, the Ministry of Foreign Affairs, the Ministry of Defence. Individually, every official who received us was very kind and as helpful as he felt he could be, but no one would approve our forms without adding a proviso that someone else in another ministry should also sanction his decision.

We had, however, one charming and sympathetic ally, a somewhat lachrymose lady in the Ministry of Information who spoke exceedingly good English. Unfortunately, we did not meet her until we had been wrestling with our problems for nearly a week. We were sent to her originally in order to have a particular rubber stamp banged on to one of our permits. We queued for an hour to reach her desk. She glanced cursorily at our papers and performed the rite. Then she read the permit in more detail, wearily removed her glasses and smiled wanly at us.

'Why do you want this?'

'We are from England, and we have come to make a film. We hope to travel through Java, Bali, Borneo and eventually to the island of Komodo photographing and collecting animals.'

The smile which had spread over her face at the word 'film' faded as I mentioned 'travel' and had disappeared by the time I said 'animals'.

'*Aduh,*' she said sadly, 'I do not think that is possible. However,' she added brightening, 'I will arrange everything for you. You will go to the Borobudur,' and she pointed to a travel poster stuck to the wall above her head showing the great Buddhist temple of central Java.

'*Njonja,*' I said, practising the formal Indonesian mode of address for a married woman, 'it is very beautiful, but we have come to Indonesia to make films of animals, not temples.'

She looked astonished.

'*Everyone,*' she said severely, 'films the Borobudur.'

'Maybe. But we film animals.'

She took the papers she had just stamped and mournfully tore them in half.

'I think,' she said, 'it is better you start again. Come back in one week's time.'

'But we can come back tomorrow and we have little time to spend in Jakarta.'

'Tomorrow,' she replied, 'is Labaran, our great Muslim festival. It is the beginning of a public holiday.'

'Does it last a whole week?' Charles asked with ill-disguised impatience.

'No. But the day after it ends is Whitsuntide and also a holiday.'

'But surely,' I said, 'this is a Muslim country, not a Christian one. You cannot have all the holidays of every religion.'

It was the only time, in all our weeks of negotiations, that we saw her in the least aggressive.

'Why not?' she said fiercely. 'When we won our freedom we told our President that we wished for *all* holidays and he has granted them.'

At the end of another week, though we had solved some of the problems which originally faced us, other bigger ones had materialized and I had to fly down to Surabaya in eastern Java to seek further permits while Charles waged a solo battle in Jakarta. When I came back I found that our friend in the Ministry of Information had really got to grips with our problems. Charles had just completed in octuplicate the largest form we had seen so far, each copy of which was furnished with specially taken photographs of his profile and full face, a complete set of his fingerprints and several high-value stamps. It had taken Charles three days of queueing to get it completed and he was understandably pleased with it. I was a little hurt.

'*Njonja*,' I said, 'perhaps I too should have such forms.'

'No, no. It is not really necessary. It is, as you might say, a luxury. I just get it for Tuan Lagus as he had nothing to do while you were away.'

After we had spent three weeks of our precious time in Jakarta we seemed no nearer to getting all the official sanctions we theoretically required than when we first arrived. I decided to confide in our ally.

'Tomorrow,' I said conspiratorially, 'we must leave. We cannot waste more time in offices. We can wait no longer.'

'Splendid,' she said, 'I think you are right. I will arrange for you to go to the Borobudur.'

'Njonja,' I said, 'please, for the last time, we are zoologists. We look for animals. We will *not* go to Borobudur.'

Standing in front of Borobudur, we were glad we had succumbed to *njonja*'s persuasions. She had been so insistent that towards the end we felt tempted to take her advice, if only to escape from the bureaucratic frustrations of Jakarta. Our resistance finally crumbled when she told us that great celebrations were about to be held there in honour of the 2,500th anniversary of the birth of Buddha. At least, we reasoned, the temple lay on our path eastward towards Komodo, and if we found that we did indeed require more permits than we had, then perhaps we could obtain them in the provincial towns. For the moment, however, our future plans were driven from our minds by the tremendous visual impact of the great temple before us. Shrines, niches and stupas rose, tier upon tier, in a magnificently rich pyramid, capping and enclosing the crest of the hill and culminating in one gigantic bell-shaped stupa, many times the size of any other beneath it, its terminal spike pointing to the sky. Behind and below, the distant plains of Java lay green with rice and palms, and beyond, on the horizon, the blue cone of a volcano trailed a plume of smoke across the turquoise sky.

Each of the four sides of the temple was pierced by an entrance at the base. As we mounted the steps on the east, passing beneath an archway and the baleful eyes of a grotesque mask which it supported, we realized that the temple, which from afar had appeared as a solid pyramid of masonry, concealed a series of high-walled corridors running round the margin of each terrace, each open to the sky and flanked on the outside by a high balustrade. Both sides of these galleries were lined with delicately carved friezes of great beauty and ornamented with sculptured flowers, trees, vases and ribbons. In niches above, figures of the Buddha sat cross-legged, gesturing symbolically with their hands, their gaze lowered in meditation. The walls of the galleries rose so high that we were enveloped and overpowered by the mass of sculptured stone soaring above and around us.

Slowly we circuited each terrace and ascended the steps to the one above, our sandalled feet echoing along the worn stone flags of the corridors. On each side of the pyramid, the Buddhas sat in distinctive poses. On the east, their hands touched the earth; on the south, they were lifted in blessing; on the west they were folded in meditation; and on the north they sat with their left hands in their laps and their right lifted in a gesture of pacification. The friezes on the walls of the lowest terrace showed scenes in the early life of Buddha. He was represented with kings and courtiers, with warriors, and beautiful women. In the background and corners of many of the panels, the sculptors had completed their compositions with charming animal studies – peacocks and parrots, monkeys, squirrels, deer and elephants. The sculptures on the terraces above, however, became increasingly austere, purged of worldly scenes and more and more preoccupied with hieratic representations of Buddha preaching and in prayer.

We climbed the steps from the fifth and last terrace and emerged on to the first three circular platforms. The mood was now entirely changed. The friezes had gone. The angular corners of the galleries orientated to points of the compass were replaced by the smooth unending sweep of the circular terraces. We had left the comparative darkness of the deep corridors and stepped out into an airy spaciousness. Seventy-two stone bells stood in circles around the gigantic central stupa. Each bell was hollow, its sides pierced into a lattice; each enclosed and half concealed a Buddha in a posture symbolizing teaching. In one place only, the bell was missing leaving the Buddha exposed and unprotected. On the last and highest platform, the final stupa rose towards the sky, vast, smooth and featureless, the very centre, core and zenith of the entire temple.

Borobudur, perhaps the finest example of the most refined phase of Buddhist architecture, was built in the middle of the eighth century. In every detail of its structure it symbolizes in stone the pattern of the Buddhist universe. The lowest terrace of all we had not seen, for it lies buried beyond the margin of the present-day base. Some have said that the builders of the temple had been compelled to cover it to prevent it being crushed and pushed outwards by the enormous weight of stone which they piled above. But the buried reliefs, when they were excavated part by part, were found to represent hellish scenes of passion and strife, and now it is thought that this terrace was deliberately placed underground as part of the symbolic design of the temple. Thus, before the Buddhist pilgrim can enter the temple, he must, in a like manner, bury and suppress his worldly passions and desires. Then as he mounts the steps spiralling the monument along the galleries, he symbolically re-enacts the life of the Buddha, purging his spirit from the dross of worldly things, ascending towards the other world until, as he reaches the upper terraces, he approaches the spiritual simplicity, the ultimate unity of the last great stupa.

Borobudur was completed only a short time before Buddhism was displaced as the official religion of the country by the Hindu faith. Six hundred years later, Hinduism in turn was driven from Java, taking refuge in the island of Bali where it still survives, and the country became Muslim to which religion it still adheres. So today, the giant temple broods on its hill, deserted by pilgrims, in an alien land. But though Buddhism in Java is now almost dead, nonetheless Borobudur still exerts a power and a presence which all who wander along its galleries must feel. The country-folk living around it still regard it with reverence; the single unprotected Buddha on the upper terraces is still visited; a simple basket of offerings lies in front of the figure and its hands are always piled with petals.

The exposed Buddha on the top terrace of Borobudur

The anniversary ceremonies were due to take place that night. As evening approached, a noisy crowd made their way to the upper terraces and assembled round the single unprotected Buddha. Soon two yellow-robed monks with shaven heads appeared and, standing by the Buddha, began a heated discussion. One of the crowd told us that the senior monk had come from Thailand specially to conduct the ceremony and that now he was arguing with the other as to the precise form the proceedings should take. Eventually, a half-hearted procession, led by the chanting monks, left on a circuit of the terrace. The people in the crowd produced bottles of water and set them in ranks at the foot of the Buddha. The crowd swarmed irreverently over the stupas, sitting on the bells, perching on the terminal spires, chattering and laughing. An Indonesian film cameraman began shouting impatiently, trying to clear people from in front of his lens to give him a clear view of the Buddha. Flash guns popped. One of the monks, enraged, began yelling to crowds to come down from the sacred stupas, but without effect. A circle of more pious onlookers at the base of the Buddha continued to chant. The other priest, who had been sitting cross-legged in meditation, got to his feet and addressed the crowd heatedly. I asked one of the onlookers what he was saying.

'First,' he replied, 'he tells us of the life of Buddha. Now he is asking who has a car to take him back to the town.'

The worshippers were due to spend the night in meditation. We stayed until nearly midnight with the noisy crowd. In the pale circle of light shed by the paraffin lamps, the lonely Buddha sat remote and withdrawn, a litter of burnt-out incense and ranks of cheap mineral water bottles at its feet. The crowd jostled, giggled and chattered. We left them to their noisy meditation.

II

The Faithful Jeep

O ur visit to Borobudur had provided us with an escape from
the enmeshing restrictions of Jakarta officialdom and
now we felt free to begin our wanderings through Java in search
of animals. Our first essential was a car and in order to hire one
we took a train to Surabaya, the largest town in eastern Java.
When we got there, we discovered that cars were virtually
unobtainable. It seemed as though we might be doomed once
again to weeks of frustration, when a great stroke of luck befell
us. We met Daan and Peggy Hubrecht in a Chinese restaurant.
Over birds' nest soup and fried crabs' claws we discovered that
Daan, who spoke Dutch, English and Malay with equal fluency,
was born in England of Dutch parents, that he managed two
sugar factories a few miles outside Surabaya, and that he had a
passion for sailing ships, sharpening knives, Oriental music and
expeditions such as ours. His immediate reaction on hearing our
plans was to insist that we left our hotel and moved into his
house, which, he said, must serve from now on as our base camp.
His wife Peggy supported this plan equally vehemently and the
next day she seemed positively delighted to have two extra
mouths to feed and her whole house littered with cameras,
recording machines, stacks of film and piles of dirty clothes.

That evening Daan produced a map, sailing schedules and timetables, and helped us draw up a detailed plan of campaign. The
eastern end of Java, he told us, was comparatively sparsely populated,
and there were many pockets of thick forest where we might find
the animals we were seeking. Further, a ferry ran regularly between

Banjuwangi, a small town on the farthermost tip of Java, and the magical island of Bali only two miles away. Then he consulted his timetables. In five weeks' time, a cargo boat was leaving Surabaya for Borneo. If we could complete our Javanese and Balinese travels in that time, he would reserve berths for us so that we could continue immediately to Borneo on the next stage of our expedition. Only one problem remained – the finding of a car – but this too Daan thought he might be able to solve for us.

'We've got an old battered jeep at the factory,' he said. 'I'll see if we can resuscitate it.'

Only two days later, the jeep was driven up outside the Hubrecht house, newly oiled, greased and fully overhauled. The next morning, we rose at five, loaded all our gear into the back, thanked the Hubrechts for all they had done for us and drove off eastwards towards we knew not what.

The jeep performed magnificently. In her way, she was a mechanical curiosity, for she was a composite creature made up of a great number of independent parts from totally different vehicles of many makes and brands. Several of the dials on her dashboard were missing and some of those which were there were not serving the purposes for which they were originally designed. The lettering and calibrations on the voltage-control meter, for example, showed that originally it had formed part of an air-conditioning plant. Her horn was sounded by brushing the naked end of a piece of flex on to a part of the steering column which had been scraped clean of paint and dirt to facilitate a good connection. Although this arrangement worked perfectly efficiently, it had disadvantages, in that each time we used it we received a minor electric shock. The tyres were of different makes and also of slightly different sizes, but all were alike in one feature – they were totally innocent of any tread, completely smooth except for one or two places where the canvas showed through in a white patch. But she was obviously a machine of great heart and stamina and we bowled merrily down the road, singing at the tops of our voices.

It was a bright sunny morning. On the horizon to our right stretched a line of volcanoes, part of the mountainous spine of Java. Close by the road, peasants in enormous conical hats stooped knee-deep in the muddy water of the terraced paddy fields, planting out young rice seedlings. Around them flocks of white egrets dabbled in the mud. Beyond, the young leaves of the widely spaced rice plants merged into a haze hanging above the brown water which mirrored the clouds, the volcanoes and the blue sky.

The road ran straight, though unevenly, along an avenue of acacia trees. Occasionally we passed a creaking ox-wagon with huge wooden wheels, driven by a turbanned peasant, and sometimes we had to swing abruptly across the road to avoid a wide carpet of rice grains which the local people had spread in a neat rectangle on the road to dry. We drove through many small villages, looking with mild curiosity at the road signs which were quite unlike any we had seen anywhere else. If there had been heavy traffic on the road, our ignorance of their meaning would perhaps have alarmed us, but as we were often the only car in sight, we were not unduly perturbed.

We had been driving for about five hours without incident, when suddenly, as we entered a village, a small policeman with a large revolver at his waist leapt into the road in front of us waving his arms and furiously blowing his whistle. We stopped. He thrust his head through the open window and harangued us in rapid Indonesian.

'We are extremely sorry, constable,' Charles replied in English, 'but you see we speak no Indonesian. We are English. Have we broken the law?'

The policeman continued yelling. We produced our passports, but these seemed to anger him even more.

'*Kantor Polisi. Polisi! Polisi!*' he shouted.

We interpreted this as meaning that he wished us to accompany him to the police station.

We were shown into a bare whitewashed room in which eight khaki-uniformed policemen were sitting gloomily around a deal

Jack Lester with a rainbow boa in Guyana.

Charles Lagus filming a column of ants on our first trip to Sierra Leone.

Recording the sound of frogs in Sierra Leone.

Examining the rock paintings near the Upper Mazaruni River in Guyana.

Inside the hut at Pipilipai, which we shared with a family of ten.

The three-toed sloth, which we soon realised had a baby hidden in her left armpit.

The tamandua, or tree anteater, above an ants' nest.

A masked actor dances accompanied by a *gamelan* orchestra, Bali.

table piled high with papers and rubber stamps. In the centre, with an even larger pistol and wearing two silver bars on his epaulettes, sat the senior officer. We apologized again for our inability to speak Indonesian and produced every letter, permit, pass and visa we could find. The officer scowled and began leafing through them. He paid only fleeting attention to our passports, ignored the numerous sets of Charles's fingerprints and finally selected one letter from the sheaf which he proceeded to read in detail. We could see that it was one from the Director of the Zoological Society of London introducing us to an authority in Singapore. 'Any help,' he ended, 'which can be given to the bearer in the maintenance of animals will be greatly appreciated by this Society.' The officer furrowed his brows, examined the signature carefully and scrutinized the copperplate heading. Charles and I handed round cigarettes to the constables sitting on the benches beside us and smiled a little nervously.

The officer carefully stacked our papers in a pile. With great deliberation he twirled the end of a cigarette between his lips, lit it, leaned back in his chair and blew a cloud of smoke towards the ceiling. Suddenly he came to a decision. He stood up, and said something gruffly which we did not understand. The constable who had arrested us motioned us to the door.

'Cells, I expect,' said Charles. 'I wish I knew what we had done wrong!'

'I have a dreadful suspicion,' I replied, 'that we were travelling in the wrong direction down a one-way road.' We followed the constable to the place where we had parked our jeep.

He motioned us to get inside.

'*Selamat djalan*,' he said. 'Peace on your journey.'

Charles shook him warmly by the hand.

'You know, constable,' he said, 'that is extraordinarily nice of you.'

And we meant it, for it was.

We drove into Banjuwangi late that evening. In the years before the war the little town had been a busy one, for the ferry running from its harbour was one of the main routes for traffic between Java and Bali. The advent of air transport had lessened its importance, but the town still retained the badges of its original rank – petrol pumps, cinemas and offices. Its single hotel, though importantly situated in the central square, was seedy and unkempt. We were shown into a small dank concrete cell, smelling oppressively of mould, its walls covered in powdering whitewash. Over each of the two beds stood an erection of wood and metal gauze resembling an oversized meat-safe, which was designed to keep its occupants safe from mosquitoes. It was so small and claustrophobic that had there been sufficient space in the room outside it I would have much preferred to sleep without its protection.

In accordance with the law and the promises we had given in Jakarta, we registered our presence the next day with the local police, the Forestry Department and the Ministry of Information. The Information Officer was a little alarmed at learning that we proposed to roam abroad in the surrounding countryside looking for animals. As he was unable to dissuade us from this course, he insisted kindly that we must take with us an official from his office to act as guide and interpreter. We had no alternative but to accept.

The guide proved to be a lanky lugubrious youth named Jusuf, who viewed the prospect of a week's wandering through the small kampongs along the coast with considerable dismay. Nevertheless, the next morning he arrived at the hotel, immaculate in white ducks and carrying an enormous suitcase, to tell us, with a slightly martyred air, that he was ready to set off into what he termed 'the jungle'. And so, with Charles at the wheel, Jusuf in the passenger seat and me, sitting between them, with my legs entangled with the gear lever, we drove out of Banjuwangi, heading for a place on the map where Daan had suggested there might be interesting forest. We stopped at each little village we passed through and, with the help of a dictionary and Jusuf, inquired about the abundance of wild life in the neighbourhood. As the day wore on, the roads became increasingly bad, the villages more widely scattered, and the country wilder and more mountainous. In the early evening we reached the top of a steep pass. As our bonnet dipped over the summit, we gasped with surprise and delight, for at the foot of the forest-clad mountain, three hundred feet below us, lay a broad curving bay, its margin bordered with groves of palm trees and its surface ridged with creamy breakers which rolled steadily towards the land until they thundered on to the white coral-sand beach. We had reached the Indian Ocean. Directly below, the yellow lights of a tiny kampong winked up at us through the dusk.

Digging the jeep out from a bog

In the days that followed we spent a great deal of our time wandering in the forest which lay behind the kampong. During the middle hours of the day the forest seemed totally lifeless, filled only with the high-pitched shrilling drone of insects. It was oppressively hot and humid, a tangle of sharp spikes, knotted creepers and occasionally a dangling orchid. Walking through it at this time of day could be an eerie experience, like walking through a town at dead of night when the streets are empty of people but strewn with litter, the trifling but evocative incidentals of human activity. So, in the forest at noon, we would find a feather, a footmark, a few hairs clinging round the mouth of a hole, the gnawed rinds of a fruit lying rotting on the floor, and we would realize that within a few yards of us many animals must be slumbering in concealment.

In the early morning, however, the forest was full of life. Many of the nocturnal animals were still abroad and the diurnal ones were beginning to wake up and feed. But this period of activity lasted only a few hours, for by the time the sun was high and hot, the daytime creatures were dozing after their meals, contented and replete, and the animals of the night had long since disappeared into their holes and burrows.

Jusuf did not come with us on these walks. He was not happy, he said, living 'in the jungle'. After a week, a Chinese rubber planter drove into the village on his way to Banjuwangi. Jusuf itched to return with him. We expressed sorrow but not desolation, so without more encouragement Jusuf packed his gear with alacrity, joined the planter in his jeep and drove off leaving us alone.

After the initial declaration of our interest in animals, I feared the villagers were disappointed that we had not produced rifles and begun hunting tigers. They were certainly mystified when they found that we spent our days observing such common and undramatic creatures as ants and little lizards. Every day, however, an old man came to see us. Sometimes he brought a small lizard or a centipede with him. Once he produced a bowl full of puffer fish, each furiously inflating itself into a creamy-coloured ball.

Two days before we were due to leave the village, he marched jubilantly up to our hut at the head of a small delegation.

'*Selamat pagi*,' I said. 'Peace on the morning.'

In reply, he pushed forward a young boy who spoke to us in Malay. Laboriously we discovered that the lad had been gathering rattan cane in the forest the day before when he had seen an enormous snake.

'*Besar*,' said the boy. 'Big. Big.'

To demonstrate the dimensions of this monster, he drew a line with his toe in the dust of the floor, took six long paces away from it and drew another line. '*Besar*,' he repeated, pointing from one line to the other.

We agreed.

There are only two snakes found in Java which attain such a size and both are pythons. The Indian python grows to a length of twenty-five feet and the reticulate python even longer, one monstrous example having been recorded as being thirty-two feet long, a measurement which qualified it as being the largest snake in the world. If the snake the boy had seen was indeed eighteen feet long, it would be a formidable creature to tackle, for a man caught in its coils would almost certainly be squeezed to death. I recalled promising the London Zoo that if we could catch a 'nice big python' we would do so.

The accepted method of capturing such monsters is simple and comparatively foolproof. It requires a minimum of three men and, for preference, the recipe recommends one man for each yard of snake. This body of eager and intrepid hunters should stand at a distance from the snake while the leader allocates duties. One man must be made responsible for the head, one for the tail, and the rest for the intervening coils. Then, on a word of command, every man leaps at the snake and grabs the section for which he is responsible. For complete success, it is important that at least the head-man and the tail-man should grab simultaneously, for if the snake has one free end, it is able to wrap itself round the man dealing with the other end, and begin squeezing. It follows, therefore, that a vital ingredient of the recipe is complete mutual confidence between all members of the team.

I looked at the turbanned group in front of me with a certain amount of misgiving. I was not doubtful of their courage individually, but I was very unsure of my ability to convey to them, without any possibility of a misunderstanding, my plan of campaign.

I talked for a long time. I drew patterns in the dust. At the end of a quarter of an hour I had succeeded in explaining enough of my plan to convince five of the men that they would have no part of it. There remained only the old man and the boy. Charles was duty-bound to be filming the operation. I proposed that the old man should leap for the tail, the boy for the middle and I would be responsible for the head. It might seem from this that I was claiming the most hazardous task, but in fact it was the job I much preferred. Although I ran the risk of being bitten, a python's fangs are not poisonous and cannot inflict anything worse than a bad scratch. On the other hand, the man dealing with the tail is liable to have a rather more unpleasant time, for a snake when it is being attacked nearly always ejects from its rear a large quantity of particularly foul-smelling excreta.

As far as I could gather, both the old man and the boy understood the plan and had agreed to help, so we gathered together our equipment and set off into the forest. The boy walked ahead cutting a path through the dense undergrowth with his *parang*. I walked behind him with a large sack and a rope. Behind me came the old man carrying some of the photographic equipment, and Charles brought up the rear with his camera, ready loaded in his hand. It would be untruthful to pretend that I was not a little nervous. Although I have an intense dislike of handling poisonous snakes, where one miscalculation can mean weeks of extreme agony and possibly death, I have a considerable affection for the non-poisonous pythons and boas. But I had never tackled one larger than four feet long and I was not overwhelmingly confident that the men who were to help me had anything but the dimmest idea of what I expected them to do when my plan went into operation and I shouted '*Mendjalankan*'. I had culled this word from my dictionary which assured me that it meant 'make go: execute'; I devoutly hoped that it was right.

Soon the ground steepened. We climbed past clumps of bamboo. Black dust and brittle fragments of dry leaves showered down on us and stuck to our sweating bodies as we forced our way between the creaking stems. As we passed through a clearing I had a sudden glimpse over the crowns of the trees growing on the slope below us and across the broad sweep of the bay to the kampong, a mile away. The boy ahead stopped and pointed to the ground. Rusting tags of iron wire and the sharp corner of a broken concrete block projected incongruously from beneath a carpet of leafy creepers. We stepped over it and found beyond a deep concrete-lined pit almost entirely concealed beneath the vegetation. Nearby a trench contoured the hill. I was reminded of pictures of the ancient monuments of Central America and Indochina which were discovered deserted and overwhelmed by the forest.

The boy spoke.

'Boom! Boom!' he said. '*Besar. Orang Djepang.*' We had stumbled upon the remains of a gun emplacement built only thirteen years earlier by the Japanese when they had invaded and occupied the whole of Java.

We walked onwards, higher up the side of the hill. At last the boy stopped. He had seen the snake near here, he said. We dumped our equipment and each of us took a different path through the bush, searching for the creature. It seemed a hopeless task. As I looked up into the maze of lianas entangling the branches of the trees, I doubted whether I should have seen the snake even if it had been in front of my eyes. Suddenly I heard the old man calling excitedly. As quickly as I could, I rushed to him. He was standing at the foot of a small tree in a clearing.

As I reached him he pointed into the branches above. Looped over one of the boughs I saw the glistening flank of a giant snake. But this was all I could see; among the confusing dapples of light and shade, of luxuriant leaves and interlacing creepers, I could not distinguish either its head or its tail. This was inconvenient: my snake-catching recipe made no mention of how to deal with giant snakes when they were in trees. I was quite certain, however, that the snake would be a better climber than I, and it was no part of my plan to have a wrestling match with one in a tree. The only solution was to get it out of the tree and on to the ground so that we could put our carefully arranged plan into action. With my *parang* in my hand, I swung myself up into the tree. The branch around which the snake had draped itself was about thirty feet above the ground. As I approached it, I saw to my relief that the reptile was lying at least ten feet along it, away from the trunk. Its flat triangular head rested on one of its enormous coils, looking straight at me with its yellow button-like eyes. It was a beautiful creature, its smooth polished body richly patterned in black, brown and yellow. It was difficult to judge its length, but the largest coil I could see was at least a foot in girth. I braced my back against the trunk behind me, and with hurried blows of my *parang* began cutting through the base of the branch.

The monster continued staring at me with a steadfast unblinking gaze. As the branch shook beneath my *parang* the reptile lifted its head, hissed and flickered its long black tongue. One of its coils began slithering smoothly over the branch. I redoubled my efforts. The bough creaked and slowly hinged downwards. With two more blows, it fell clear, carrying the python with it, and landed with a crash close by the boy and the old man.

'*Mendjalankan!*' I roared. 'Make go: execute.'

They gaped at me uncomprehendingly.

I saw the snake's head appear from between the leaves of the fallen branch and it began sliding out, heading for a clump of bamboo on the other side of the clearing. If it reached it and succeeded in entwining itself among the massive bases of the bamboo stems, we should never catch it.

I began scrambling down the tree, as fast as I could. '*Mendjalankan!*' I bawled in exasperation at my team who were standing beside Charles and his camera, watching dumbfounded.

With a final jump, I landed on the ground, seized the sack and ran after the snake which was now within three yards of the bamboo. If we were going to catch it, I should have to tackle it myself. Fortunately, it was so intent on reaching the bamboo that it paid no regard whatsoever to me as I ran after it but continued wriggling onwards with surprising rapidity for so large a snake.

I caught up with it just before its head entered the bamboo. I snatched its tail and jerked it backwards. Infuriated by this indignity, it turned on me, opened its mouth and drew its head back in a striking position, its black tongue flickering in and out. I took the sack in my right hand and threw it, like a fisherman casting a net, so that it dropped neatly over its head.

'Hoop-la,' yelled Charles from behind his camera.

I pounced on the sack, and fumbling in the folds gripped the snake by the scruff of its neck. Then quickly, remembering the recipe, I grabbed its tail with my other hand. I stood up in triumph. The great snake twisted and struggled, coiling itself into loops. Its body, which I estimated was at least twelve feet long, was so heavy and cumbersome that though I raised its head and tail above my head, its middle coils still lay on the ground.

It was at this moment, as I held the snake aloft, that the boy at last decided to come to my assistance. He arrived just in time to receive a jet of foul fluid all over his sarong. The old man sat down and laughed until tears ran down his cheeks.

Although we spent most of our time working close to the kampong, occasionally we travelled farther along the coast to visit other villages and explore new parts of the forest. To do this, we had to drive our jeep over roads surfaced with savagely angular boulders, across fords so deep that the water lapped well above the hubs, and through quagmires of such soft mud that on several occasions the jeep sank down with whizzing impotent wheels until the crankshaft and axles rested flat on the surface of the bog.

Some people automatically bestow a name and a sex upon every car they drive. I had always considered such an attitude to a piece of machinery to be somewhat sentimental, but I changed my views during these journeys. There was no doubt whatsoever that our jeep had a strong and highly individual personality. She could be both capricious and temperamental, yet always she was extremely loyal. Often in the mornings, when we were alone and unobserved, she would refuse to start unless we cajoled her with a great deal of cranking. But if we were being watched closely by a party of villagers or if we were visiting some local official and it was essential that we should make a dignified departure, then always she shook into life at the first touch of the starter. Once in action, she was a machine of great courage. Never once did she jib at any of the obstacles with which we faced her.

This is not to say that she was not ageing and, in some respects, infirm. Once, she developed a persistent slow leak in one of the pipes carrying hydraulic fluid to her brake drums. We hardly ever used the brakes as they were appallingly uneven and whenever we applied them she slewed alarmingly across the road, but the prospect of losing all the irreplaceable hydraulic fluid and with it the jeep's entire braking power was so serious that we decided it must be remedied. The only cure we could devise was a brutal one. We unscrewed the offending pipe, and with two boulders hammered it solid. To our surprise, she braked very much more evenly after this surgical operation than before it.

On one occasion, she demonstrated her loyalty in an even more positive way; she came to our aid during one of our regular skirmishes with our mechanical arch-enemy, the tape recorder.

This was a machine of extreme ill-temper, which, having been built for better things, plainly resented performing the odd functions we required of it. Often, having tested it to make sure that it was in full working order, we would set up our microphone and sit for hours waiting for a particular birdcall. The bird would sing, we would switch on the machine in triumph, only to find that the spools refused to revolve, or, if they did turn, that the circuits inside were not functioning properly. Usually, after this show of petulance – and after the bird had disappeared – the machine would miraculously cure itself and behave perfectly for the rest of the day. If, however, it was stubborn, we had two methods of dealing with it. The first was to smack it very hard. Often this was sufficient but if such treatment failed we took more drastic action. First we dismantled it as far as possible and spread out its valves and other components in neat rows on a banana leaf or some other conveniently smooth surface. We rarely found anything wrong, but this did not matter; we simply reassembled everything in exactly the same manner as before, and the machine would work like a charm.

The occasion on which the jeep aided us in one of these battles was the only time when we discovered something organically amiss with the recorder's insides. It was a particularly embarrassing moment, for the entire village had assembled to sing for us. I switched on the recorder with a flourish, but the microphone remained quite dead. Blows having failed, I began to dismember it using the tip of my *parang* as a screwdriver. To my surprise, I found that one of the internal wires had, in some mysterious way, been broken. Furthermore, this particular wire was not long enough to enable us to overlap the two broken ends and twist them together. We had no spare wire with us. I looked up to apologize to the headman and cancel the concert when my eye fell on the jeep, which was parked nearby. Beneath her front axle dangled something I had not seen before: a long yellow wire. I walked over and examined it. I could not see what it was attached to but the lower end hung quite free. I cut off six inches with my *parang*. Fitting it to the tape recorder was not easy, but having done it and reassembled the machine, I found that it worked perfectly. I wondered if perhaps the recorder had been shamed into good behaviour by this act of self-sacrifice on the part of the jeep.

These experiences had given us such confidence in our car that when the time came for us to say goodbye to our friends and leave the kampong we had no qualms whatever about her ability to carry us and our equipment back to Banjuwangi and on to Bali. But we had not driven for more than an hour when suddenly she began to stagger and shake, her near-side front wheel vibrating distressingly. We stopped and Charles crawled beneath her to investigate. He emerged, oil stained and dirty, with bad news. The four bolts connecting the steering rods with the front wheel had finally given way under the strain of careering over the execrable roads. Each of them was sheared in two.

The situation was serious. We could not go on, for we would be unable to steer her round the next corner. The nearest village was ten miles away and as far as we knew the nearest garage was in Banjuwangi. It was then that our remarkable machine once again demonstrated her resourcefulness. As Charles sat with the oily pieces of shattered metal in his hands, he noticed a line of bolts of a similar calibre in the underpart of the chassis. He unscrewed four of them. As far as we could see, they did not seem to serve any particular function and the jeep showed no reaction at their loss. He crawled beneath the front axle with them in his hand. After a great deal of grunting and hammering, he reappeared smiling. They had fitted exactly. We restarted the engine and moved off. Gingerly we negotiated the next corner. On we went with increasing confidence and finally late that evening we drove into Banjuwangi at full speed. Ahead of us lay Bali and many miles of roads as bad as any we had crossed so far, but this was the last time that our aged but wonderful machine was to complain of the savage treatment to which she was being subjected.

12

Bali

———

The explanation of many of the characteristics which make Bali so different from its neighbouring islands is to be found in its history. A thousand years ago, Hindu kings ruled over Java, Sumatra, Malaya and Indochina. Their capital was in Java, and, as their power waxed and waned, so Bali was either a vassal of Java or an independent state. In the fifteenth century, the islands were governed by emperors of the Modjopahit dynasty. Towards the end of their reign, Mohammedan missionaries began spreading a new faith in Java. Soon local princelings became converted to Islam and proclaimed themselves independent of the Hindu Modjopahits. Civil war overtook the islands and the last emperor, according to one account, was told by his priest that at the end of forty days the rule of the Modjopahits would be over. On the fortieth day, the emperor gave orders for his supporters to burn him alive. His young son, the prince, fearful of the fanaticism of his Muslim countrymen, fled to Bali, his last remaining colony, taking with him his entire court. The transfusion of the finest of Java's musicians, dancers, painters and sculptors which Bali received by this migration has profoundly affected the character of the island and is perhaps one of the reasons why today the people of Bali are so extraordinarily gifted in the arts. Certainly, the Hindu faith which the Balinese, alone among the peoples of Indonesia, still follow so pervades the island that nearly every aspect of their existence is governed and modified by it, from the design of their villages to the mode of their dress and their everyday behaviour. Furthermore, Balinese

Hinduism, isolated as it is from the parent faith in India by a barrier of Islam, has evolved into a highly idiosyncratic version of the original belief so that Bali is virtually the possessor of a unique religion.

As we drove through the lovely villages and fertile fields, past plantations of palms and bananas, we, as all visitors to Bali are supposed to do, felt that we had arrived in an island paradise, the embodiment of everyone's dreams of the ideal tropical island, where the people are beautiful and peace-loving, where the ground is so rich that fruit-bearing trees grow with the abundance of weeds, where the sun never ceases to shine and where man at last is in harmony with a bountiful and beneficent nature.

It was fortunate for us that we had approached the island by this route. Many visitors are compelled to arrive by air at Denpasar, Bali's largest town, and Denpasar, as we discovered when we drove into it late that night, is far from being a typical sample of any island paradise. It is dominated by cinemas, cars, enormous hotels, souvenir shops and a concrete dancing arena outside the main hotel where visitors can watch specially arranged dances as they sit comfortably in their wicker chairs with a whisky and soda at their elbow.

We had come to the town for, in addition to everything else, it contained the usual proliferation of offices and we had to register our presence with nearly all of them. But here in Denpasar we were very fortunate, for we were guided through the formalities at great speed by Mas Soeprapto. We had met him within a few days of our arrival in Jakarta where he was an official of the radio station. Although he was not by birth a Balinese, he was a great expert on the island's music and dancing and had been the business manager of a group of Balinese dancers which had recently toured the world. As a result he had a shrewd appreciation of the differences between western and oriental people, and he was one of the few Indonesians we had met who realized how frustrating we found the procrastinations of officialdom. When he had volunteered to be our guide should we visit Bali we had been delighted, and now, in Denpasar, we began to realize how very fortunate we were.

I expected Mas' first action would be to take us out of the hybrid civilization of Denpasar back to rural Bali. The first night, however, he led us past the neon lights of the town centre to the home of a nobleman in one of the quieter backwaters of the town. He had brought us to see the preparations for a great feast that was to be held the next day. The pavilions of the household were thronged with people. Women were deftly constructing beautiful lacy decorations from palm leaves, pinning the component whirls and tassels into position with thin slivers of bamboo. Pyramidal rice cakes, some white, some pink, were being laid out in long rows on napkins of olive-green banana leaves. Garlands of flowers were being hung from the eaves of buildings and rich ceremonial cloths draped round the shrines of the household gods. Between the pavilions lay six turtles still alive, their fore-flippers cruelly pierced and tied with a thong of rattan cane, their dry leathery heads sunk to the ground. They blinked slowly, their weary glazed eyes weeping copiously as the laughing, chattering crowd swept round them. They would be slaughtered that evening.

The next day Mas took us back to the house. The courtyard was even more tightly packed with people than it had been on the preceding night. All were wearing their best clothes, the men in sarongs, tunics and turbans, the women in tight blouses and long skirts. The prince, the head of the household, sat cross-legged on a small platform chattering to the more important guests, drinking small cups of coffee and eating gobbets of turtle meat spitted on bamboo sticks. In front of him a boy sat playing a dulcimer-like instrument with five bronze keys which he struck with a mallet, producing a monotonous tinkling tune.

Mas told us that the feast was to celebrate the performance of a tooth-filing ceremony. Jagged uneven teeth are considered by the Balinese to be characteristic of beasts and demons. Consequently, when every man and woman comes of age he should have his teeth filed so that all irregularities are removed and they are smooth and straight. The ceremony nowadays is not as widely practised as it was, yet even now, if a person dies without having submitted to the ritual, his relatives will file the teeth of the corpse before its cremation lest the bestial attributes of uneven teeth should deny him entrance to the world of spirits.

Towards midday, a small procession emerged from the family pavilion. It was headed by the initiate-to-be, a young girl. Her torso was tightly wrapped in a red cloth richly painted with gold floral patterns. She carried over one shoulder a long sash of a similar material and on her head she wore a splendid and elaborate crown of gold leaf and frangipani flowers. Several older women, less extravagantly costumed, accompanied her, chanting as they walked. The column advanced down the crowded alleys to a pavilion hung with batik cloths. On its steps, a white-clad priest awaited her. She stopped in front of him and held out her hands. With hieratic gestures he took a funnel of woven bamboo through which he poured water so that it trickled over her outstretched fingers. His lips moved as he performed the ceremony, but the dulcimer player and the chanting women drowned his words. The priest put the funnel aside and led the girl into the pavilion. There she lay down on a couch, her head resting on a long pillow covered in a specially woven cloth of great magical significance. The priest blessed his instruments and leaned over her to begin the work of filing. The escorting women sang more loudly. One of them held her feet, and two others her hands as she lay prostrate. If the girl cried out during the operation, we did not hear it above the chants. Every ten minutes the priest stopped and held a mirror so that she might see how the filing was progressing. After half an hour the work was finished. The girl rose and was led out of the pavilion, pausing on the steps so that all might see her. Her eyes were welling with tears, her magnificent headdress awry and bedraggled, for her singing escorts had plucked some of the gold leaves from it to wear in their hair. In her hand she held a small decorated coconut shell containing the filings which she had spat out. She walked back through the pavilions on her way to the family temple where she would bury the filings of her teeth behind the shrine of her ancestors.

The tooth-filing ceremony

When we left the town the next day, we found to our surprise that, in spite of the busy international traffic flowing through Denpasar and its airport, western influences had hardly spread beyond the boundaries of the town; we had only to abandon our jeep and walk for short distances along the narrow tracks which wound through the rice fields to discover villages which were still totally unaffected by the modern world. With Mas Soeprapto as our guide, we spent day after day wandering through the island and every day – and every night – it seemed that some entertainment, some ceremony, was being held in one of the village houses or a temple.

The Balinese are a people possessed by a passionate love of music and dancing. Every man, whether he is a prince or a poor rice-farmer, seems to have the ambition to perform in his village orchestra or dancing group, and those who are not talented enough to do so count it a privilege to subscribe what they can afford to help in the purchase of costumes or fine instruments. Even the poorest, smallest village owns, communally, a *gamelan*. This is the traditional orchestra of Bali. The majority of its instruments are metal ones – large hanging gongs, smaller ones set horizontally in racks, tiny cymbals and many different variants on the dulcimer-like instrument we had seen in the ceremony at Denpasar. In addition to these, there may be a *rebab*, the two-stringed Arab fiddle, bamboo flutes and, always, two drums.

Most of these instruments are extremely expensive. Balinese smiths are able to forge the bronze keys for the dulcimers, but

The gamelan

the secret of making the clearest-sounding and most musical gongs is possessed only by the craftsmen of a small town in southern Java and a fine gong is therefore a treasured possession, worth a great deal of money.

The music produced by the *gamelan* is of the most ravishing kind, full of subtle percussive rhythms, plangent ripples and crashing chords. I had expected that I should find it too foreign, too exotic, to give me any real pleasure. Yet it was not so. The musicians played with such verve, conviction and dedication, and their music was alternately so exciting and so tenderly contemplative, that we were enraptured by it.

The youngest members of the gamelan

Twenty or thirty people are necessary to play the full *gamelan*, and they perform with a precision and accuracy of timing which would rival that of any European orchestra. None of their intricate compositions is ever written down; the musicians carry them only in their memories. Furthermore, every orchestra's repertoire is so extensive that it is able to play for many hours on end without repeating any one composition.

This high professional skill is only gained by arduous practice. Each night as dusk fell the village musicians gathered in a pavilion to begin rehearsals. As the tinkles and sonorous crashes of the orchestra rang round the village, we, with Mas as our sponsor, sought out the rehearsal pavilion to sit and listen. The leader of the *gamelan* is always the drummer and it is through the beats of his drum that he is able to control the orchestra's tempo. Usually, however, he is an equally skilful performer on all the other instruments and he often stopped the music and walked over to one of the dulcimer players to demonstrate exactly how a theme should be played.

It was at these rehearsals that we saw for the first time the young girl dancers who perform the *legong*, one of the most beautiful and tender of all Bali's dances. None of the three dancers could have been more than six years old. The instruments of the *gamelan* were ranged round three sides of a square, and in the arena so formed, the girls took their lessons. Their teacher was an old grey-haired woman who as a young girl had been a famous *legong* dancer herself. Her method of instruction was sharply, almost savagely, to thrust her pupils' heads, arms and legs into the correct position as they danced. Hour after hour, the music continued and the children, under the severe eyes of their tutor, stamped and gyrated with quivering fingers and jerking eyes. Towards midnight the music at last came to an end. The lesson was over and in an instant the dancers changed from impassive sphinx-like figures to laughing scruffy children who ran giggling and shouting back to their homes.

13

The Animals of Bali

O ur reference books had told us that the animals of Bali
were of no great interest, that with the exception of one
or two birds, all the creatures to be found there occurred in
greater numbers in Java. But the books did not mention the
domestic animals, and when we finally settled down in a village
for the last two weeks of our stay in the island we discovered
to our delight that many of these are as peculiar to Bali as its
dances and its music.

Every morning processions of snow-white ducks waddled out
of the village. They were quite unlike any we had seen before,
possessed charming little pompoms of curly feathers on the backs
of their heads which invested them with a gay, slightly coquet-
tish air, as though they were creatures from a story book dressed
up for a carnival. Behind each flock walked a man or a boy
carrying a long slender bamboo which he held horizontally over
the heads of his charges so that the bundle of white feathers,
which was attached to its tip, bobbed up and down in front of
the leader of the parade. The ducks, from hatching, had been
trained to follow these feathers and so, guided by the lure, they
filed jauntily along the narrow paths until they reached a paddy
field which had been recently harvested or newly ploughed.
There the herdsman planted the bamboo slantingly in the mud
so that the feather bundle hung dancing in the breeze within
sight of the flock; and there the birds remained all day, happily
dabbling in the mud, never straying far from their hypnotic white
bundle. In the evening their guardian returned, took up the pole

and once again the gay quacking procession followed the bobbing feathers along the dykes and back to the village.

The cows were also decorative creatures. They had reddish-black coats, white knee-length stockings and neat white patches on their rumps, for they are the domesticated descendants of the banteng, the wild ox that is still found in the forests of South-Eastern Asia. The breed has remained so pure in Bali that many of these cows are still indistinguishable from the beautiful wild creatures which sportsmen in Java take such pains to stalk and shoot.

The origin and ancestry of the village pigs, however, was something of a mystery to us, for they resembled no other breed, either wild or domestic. When we first encountered one I thought it was monstrously deformed. Its backbone slumped between its bony shoulders and haunches, seemingly dragged down by the weight of its huge belly which sagged like a sack of sand and rubbed in the dust as the beast moved. This ugly characteristic was no individual malformation, and we soon discovered that it was shared by all the pigs of Bali.

The villages swarmed with dogs, but if they had a unique quality it was that they were by far the most loathsome curs we had ever encountered. They were all half-starved and most of them hideously diseased. Their ribs and backbones showed with pathetic clarity through their ulcer-covered skins. They existed on the garbage which they scavenged from the household refuse tips, supplemented by the tiny portions of rice which the Balinese place every day in front of the shrines, gateways and pavilions as offerings to the gods. It would be a merciful act to shoot the majority of these miserable brutes, yet the villagers allow them to breed unrestrictedly. Not only do they tolerate them during the day, but they welcome their incessant nocturnal howls for they believe that this noise frightens away the evil spirits and demons which roam through the village during the night seeking to invade the households and possess the sleeping occupants.

A large and particularly vociferous dog took up its nightly

station directly outside the pavilion in which we were supposed to sleep. On our first night there, at three o'clock in the morning, I could stand its howls no longer. I decided, on balance, that I preferred to take my chance with a demon rather than spend the rest of the night with such an unpleasant guardian, so I picked up a stone and threw it in the direction of the brute in the hope that I might persuade it to carry out its duties else-where. The only result I achieved, however, was to turn its melancholy howls into furious yaps which roused every other dog in the village and resulted in a deafening chorus which lasted until daybreak.

It seemed to us that the Balinese had little regard for the welfare of animals. Not only did they allow these walking bundles of diseased skin and bone to wander loose, but they also enthu-siastically staged fights between both crickets and cocks.

Cricket-fighting is a comparatively minor sport. The insects are kept in small cages of carved bamboo. When a contest is to take place, two small circular flat-bottomed pits are dug in the ground and a tunnel bored connecting the two. A cricket is placed in each hole and their owners, sitting beside them, irri-tate them with flicks of a quill. Eventually one of them is persuaded to crawl through the tunnel and there, infuriated and goaded by the quill, it attacks the other insect. The two battle furiously, seizing one another's legs with their jaws and rolling over and over. At last, one finally tears a limb from its opponent and is declared the winner. The maimed one is thrown away and the victor, chirruping, is replaced in its cage to fight again.

Cockfights are much more serious affairs. At certain times of the year they are ritual necessities, for the Balinese gods peri-odically require that fresh blood shall be spilled in their honour. But the fights are also great sporting occasions and huge sums of money are gambled upon the results. We heard of one man who, with unlimited faith in his fighting cock, mortgaged his house and all his belongings to place his entire wealth, a sum the equivalent of several hundred pounds, upon the success of

A cricket fight

his bird. The bet, however, was so high that no one else was able to accept it.

The main street of the village was lined by bell-shaped cages of split bamboo which contained the cockerels. All times of the day old men sat fondling their birds, clasping them by their breastbones, bouncing them up and down on the ground, ruffling their neck feathers and assessing their potentialities as murderers. Every feature of a bird has its significance – its colour, the size of its comb, the brightness of its eyes – and by such characteristics as these each owner decides the type of bird against which he wishes to match his own.

One morning there was a great deal of activity in the market place. Small stalls were being set up by *saté* sellers and by women who dispense palm wine and the chemical-pink drink of which the Balinese are so fond. Preparations were being made for a big cockfighting festival. An arena was laid out in one of the

Balinese cockfighters

large thatched shelters in which communal meetings were usually held, and the fighting ring demarcated by strips of bamboo pegged down on the mud floor. Around this were set fences of woven palm leaves a foot high beyond which the spectators would sit.

On the day of the festival, men arrived in the village from remote hamlets many miles away carrying their fighting cocks, each in a small satchel woven from a single palm leaf with a gap at the back through which the bird's tail feathers protruded. A noisy crowd gathered round the ring. The judge, an old man, seated himself cross-legged by the screen. On his left stood a bowl of water in which floated half a coconut shell, pierced at the bottom by a small hole. This was his clock which measured each fighting period in units of the time taken for water to flow into the coconut shell and sink it. At his left hand lay a small gong which he would strike to mark the beginning and end of the round.

A dozen men stepped over the screen carrying their birds. After a great deal of bouncing and feather fluffing, the birds were paired off and a fighting order decided upon. The ring was cleared and the birds taken away to have their six-inch-long, razor-sharp fighting blades tied to one of their legs in place of their natural spurs which had long since been removed. The first pair, fully armed, were brought back into the ring. Once again they were held facing one another so that they were mutually provoked to show their aggressive spirit by crowing and erecting their neck plumes. These preliminary displays enabled the crowd to judge the qualities of the birds, and the gamblers shouted bets to one another across the ring. The time-keeper struck the gong and the fight began. The birds met, beak to beak, circling one another, feathers erect. Crowing fiercely, they flew into the air, striking murderously at one another with their steel spurs, the blades flashing. One of the birds had no stomach for the fight and kept running from the ring. Each time, the crowd scattered quickly, for a stray slash from the cock's knife can inflict a very severe wound on a man as well as a bird. Again and again its owner gingerly recaptured it. Eventually a deep black stain, oozing through the feathers beneath its wing, showed that it had been seriously injured. Again it fled and again it was caught and put back to face its savage attacker. But the wounded bird would not face his opponent and no cockfight is settled until one of the birds is dead. The time-keeper struck his gong and called an instruction. One of the bell-shaped cages was placed in the ring and the two birds put beneath it. Here, unable to escape, the poor wounded creature was finally slaughtered.

The second fight was even more revolting, for both birds were brave. For round after round they fought, tearing at one another's wattles and neck feathers, striking again and again with their blades until both birds were pouring blood. Between the rounds, each owner did his best to resuscitate his bird by placing its beak in his mouth and blowing air into the cockerel's lungs. One wiped some of the blood on to his finger and made the bird taste it. Soon one of the cocks, weak from loss of blood and staggering with its wounds, received a mortal stab. It sank gasping to the ground. The victor continued snatching at the dying bird's wattles, and trying to peck at the already glazing eyes of the corpse until at last its owner pulled it away.

During that day many birds died and a great deal of money changed hands. That evening many households in the village ate chicken meat with their rice. Presumably, the gods were placated.

Our last night in Bali we had to spend in Denpasar before returning westwards to the ferry which would take us back to Java. We had rooms in a small *losmen* in the quieter part of the town, and after depositing our baggage there we spent the evening visiting and making our farewells to the officials who had helped us. We did not return until nearly midnight. The owner of the *losmen* was waiting for us, anxiously wringing his hands. It appeared that the driver of a lorry had brought down a message from the village which the innkeeper had promised to pass on to us. It was clear that it was both urgent and important but unfortunately our ignorance of the language prevented us from understanding any word of it. The innkeeper was most distressed. With his brows twisted in anguish he said with great vehemence, '*Klesih, klesih, klesih.*' We had no idea what he meant, but he was so insistent that we felt we must drive back thirty miles to the village to solve the problem. If we did not go that night, we should be compelled to leave Bali in the morning and never know what urgent affair had awaited us in the village.

It was one o'clock before we reached the village. After disturbing the sleep of several villagers we at last discovered the person who had sent us the message. He was Alit, the young son of the household in which we had been staying. Fortunately for us he spoke a little English.

'There is,' he said haltingly, 'in the next village a *klesih*.'

I asked what a *klesih* was. Alit did his best to explain. It was some sort of animal, but we could not identify it more exactly from his description. The only thing to do was to go and see it. Alit disappeared and came back holding aloft a flaming palm-leaf torch to provide us with a light and together we set off through the fields.

After an hour's walk, we saw the dim outlines of a hamlet ahead of us.

'Please,' said Alit, 'let us be polite. This is a village of bandits, very fierce men.'

As we entered the village, the watchdogs raised the alarm with choruses of howls and yaps and I fully expected the 'very fierce men' to rush out waving swords; but no one appeared. Perhaps the villagers, hearing the noise, merely assumed that the dogs were howling at yet another band of prowling evil spirits.

Alit led us through the deserted streets to a small house in the centre. He knocked loudly on the door and at last a tousle-headed man, rubbing the sleep from his eyes, opened it to us. Alit explained that we had come to see the *klesih*. The man seemed disbelieving but at last Alit convinced him and we all went inside. From underneath his bed the man produced a large wooden box tied firmly with a cord. He undid it and opened the lid. An earthy acrid smell wafted into the air. He took out a round bundle the size of a football, covered in brown triangular scales.

The *klesih* was a pangolin. He placed it gently on the floor where it lay, its sides heaving slowly up and down.

We remained quiet for a few minutes. Slowly the ball began to uncurl. First it unwound its long prehensile tail. A pointed wet nose appeared and behind it a small inquisitive face. The little creature looked around shortsightedly, blinking its bright black eyes and panting. None of us moved. Emboldened, the pangolin rolled over on to its legs and began to trundle round the room like a tiny armoured dinosaur. He reached the base of the wall and with vigorous snatches of his foreclaws, began to excavate a hole.

'*Aduh!*' said his owner, stepping over smartly and picking up the animal by the end of its tail. The pangolin curled himself into a ball again by rolling upwards like a yo-yo. The man put him back into the box.

'One hundred rupiahs,' he said.

The pangolin curls up while holding on to my hand with its prehensile tail

I shook my head. Some species of anteaters will accept minced meat, condensed milk and raw eggs as a substitute for their normal diet, but the pangolin will only live on ants and those only of the right sort. We could not therefore hope to take him back to London.

'What will the man do with the *klesih* if we do not buy it?' I asked Alit.

In reply he grinned and smacked his lips. 'Eat him,' he said. 'Very good!'

I looked at the box. The pangolin was peering out, with his forepaws and chin resting on the edge, hopefully exploring the sides for ants with his long sticky tongue.

'Twenty rupiahs,' I said firmly, excusing my extravagance to myself with the thought that at least we could take some photographs before we set the animal free. The man put the lid on the box and handed it to me with alacrity.

Alit set aflame another palm-leaf torch and holding it above his head he led us once again through the village and out into the rice fields. Holding the box firmly under one arm, I lagged a little behind him and his pool of light. The moon was full and yellow. The plumed palm trees waved gently against the black velvet sky ablaze with stars. We walked silently along narrow muddy paths, waist deep in the growing rice, the surface of which shimmered with the green unearthly incandescence of dancing fireflies. As we passed the intricate silhouette of a temple gateway, the warm air was filled with the delicate perfume of frangipani. Faintly above the shrilling of the crickets and the gurgle of the water trickling through the irrigation channels, we heard the distant booms of a *gamelan* playing for an all-night festival.

It was, we knew, our last night in Bali and we were deeply sorry.

14

Volcanoes and Pickpockets

W hen we reached Surabaya, Daan welcomed us with piles of letters from England, an endless supply of iced drinks and some very good news. Not only had he managed to reserve berths on a cargo vessel sailing to Samarinda, a small town on the east coast of Borneo, but he was also free to come with us for at least two weeks to act as interpreter. The ship, it was true, was not due to sail for five days, but far from being irritated at this delay both Charles and I were secretly delighted, for after our weeks of travelling we both felt in need of rest. As soon as we had packed all our exposed film in hermetically sealed boxes and overhauled all our equipment, Daan and Peggy took us away to a bungalow in the hills outside Surabaya.

The flat steamy plains around the town are intensively culti-vated. We drove along avenues of tamarind trees, past acres of flooded rice paddies and fields of tall waving sugar cane. As we gained height the air became cooler and crisper. Kapok planta-tions covered the hillsides, the tall trees laden with pods, many bursting with white fluffy fibres. Tretes, the village in which we were to stay, lies two thousand feet above the sea on the flanks of Walirang, one of a group of spectacular pyramidal volcanoes.

The island of Java forms part of an immense volcanic chain which runs from Sumatra south and eastwards through Java, Bali and Flores and then swings northwards to join the Philippines. The volcanoes in this arc have been responsible for some of the most violent and disastrous eruptions in historic times. In 1883, Krakatoa, a small volcanic island lying in the sea between Java and Sumatra, exploded with such titanic power that it blew away over four cubic miles of rock, covering the sea over a vast area with floating pumice and causing giant tidal waves which swept over the low-lying coasts nearby, drowning thirty-six thousand people. The noise of this stupendous eruption was heard nearly three thousand miles away in Australia.

Java alone contains a hundred and twenty-five volcanoes of which nineteen are in a state of continual activity, sometimes merely smoking and sometimes erupting in paroxysms of enormous violence as when, in 1931, Merapi burst into action killing thirteen hundred people. It is hard to minimize the influence of volcanoes on Java and its people. Their shapely but ominous cones dominate the landscape; their lava and ash, which for hundreds of centuries they have poured over the land, has decomposed to produce one of the most fertile soils in the world; and the terror of their active periods has made them, in the islanders' mythology, the homes of powerful gods.

Walirang was dormant, yet from the garden of our bungalow in Tretes I could see wisps of smoke issuing from its summit eight thousand feet above. Already the cool invigorating air had dispelled the lassitude I had felt in the heat of Surabaya and I determined to climb up the mountain and look at the crater.

Charles did not share my enthusiasm, even after I had discovered that it was possible to ride the greater part of the way, so I arranged for only one horse to be brought to the house early the next morning.

It arrived just before dawn, led by an aggressively cheerful old man. The horse was a small bony creature which stood hanging its head with a depressed expression on its face. Its owner smacked it enthusiastically on its flank and smiled widely, exposing his stumps of teeth, blackened by betel nut. He made it clear that he considered his animal to be one of the strongest and most agile of all the horses in Tretes, well worth every rupiah of the exorbitant price I had agreed to pay for its hire.

I climbed on its back feeling almost ashamed for seeking any assistance from such a pathetic creature. The hillman prodded it energetically, and at a snail's pace, with the stirrups almost touching the ground, we set off slowly through the village.

Soon the level road began to steepen. My mount looked mournfully at the rising path in front of it, broke wind expressively, and stopped. Its owner smiled and tugged ferociously at the reins. The horse refused to move. From the rumbles beneath me, it was clear that the poor creature was afflicted with appalling indigestion, and in sympathy I dismounted. Immediately it trotted nimbly up the path. After half a mile we reached a level patch and my guide indicated that it was now proper that I should ride again. At first all was well, but after ten minutes the animal slowed to a standstill. The old man once more pulled the reins, this time with such force that they broke from the bit. The horse was now virtually unsteerable and I got off. As I did so, one of the girths parted. The horse looked so dejected as its accoutrements disintegrated around it that I could not bring myself to make any further attempt to ride. Together we walked on upwards. Our progress was slow, for every half hour I had to stop to allow my expensive mount and its owner to catch up.

We were now walking through forest unlike any I had seen elsewhere. It was rich in orchids, and tree ferns grew abundantly, their crests, like giant clumps of bracken, springing from the tops of bare graceful trunks. As we climbed higher, we entered groves of casuarina trees, superficially resembling a pine forest, the trunks widely spaced and free from entangling creepers, the branches whorled with long drooping needles and bearded with dangling tufts of Spanish moss. After five hours I reached a small encampment of low thatched shelters. Propped against the turf walls stood tubular wicker baskets loaded with brilliant yellow sulphur. Several men emerged from the huts and stared at me, scowling. They were small, dark-complexioned people, barefooted, clothed in worn shirts and sarongs and wearing a variety of headgear ranging from battered felt hats and torn *pitjis* to simple turbans.

I sat down by the side of the track and began to eat some sandwiches. After a few minutes my guide arrived. He explained that the crater was now only one hour's walk away but that the path from here on became steeper and was unsuitable for his horse. He, therefore, would remain here so that he might repair the harness and the horse could recover from its exertions.

While I finished my meal, the sulphur-gatherers began talking in undertones to the old man, shooting suspicious glances in my direction. Six of them eventually picked up empty baskets and set off along a narrow track through the casuarinas. In spite of their unwelcoming looks, I attached myself to the end of their column and plodded along behind them. We left the forest and began climbing over rough boulders of lava interspersed with stunted bushes. We were now over nine thousand feet high; the air was thin and it was very cold. Periodically, mist swept up the side of the mountain and enswathed us. We walked in silence, the men ahead ignoring my presence. After half an hour one of them began to sing in a high falsetto voice. The tune was a plaintive one and the words, as far as I could understand them, were an extemporary recitative of current events.

'*Orang ini*,' he chanted, '*ada Inggeris, tidak orang Belanda.*'

This at least I understood. It meant 'This man is English not Dutch'. Since I figured in his song, I decided to try my own hand at extemporization. It took me several minutes to link together a few words from my very limited vocabulary into some sort of sense, and as soon as the singer ahead came to the end of his verse, I boldly launched out into my own song.

'This morning,' I sang, imitating his tune, 'I eat rice. Tonight I eat rice. Tomorrow, I am so sorry, I eat rice again.'

I realized that this was neither witty nor particularly topical, but its effect, nonetheless, was phenomenal. The men stopped, sat themselves on boulders and laughed until tears came to their eyes. As soon as they recovered I produced a packet of cigarettes and we all had a smoke together. We tried to converse, but I fear they understood little of what I attempted to say and I, without my dictionary, only had the vaguest idea of what they were saying to me. Nevertheless, their frigid attitude had vanished and when we got to our feet again it was as a united party.

Together, we climbed up to the summit. The main crater, a vast cliff-lined shaft running vertically into the mountain, seemed lifeless, its floor, two hundred feet below, a wilderness of boulders. But Walirang was by no means dead, for beyond the crater rose a turbulent column of white smoke. We scrambled towards it and looked down the flanks of the mountain. The smoke rose not from a single source but from a hundred small scattered vents, each roaring and hissing, jetting its fumes into the air, so that it seemed that a huge area of the mountainside had caught fire and was smouldering fiercely. The atmosphere was full of acrid sulphurous fumes which snatched at every breath I took, the ground beneath my feet was coated with a thick dusting of yellow sulphur. Through the shifting billows of smoke, I caught glimpses of tiny figures working in the heart of the inferno. They had blocked the main fumaroles with rocks so that the gas, instead of belching directly into the atmosphere, was diverted down a series of radiating pipes, cooling on its way and shedding its load of sulphur. Some of these pipes, already choked with the precious mineral, were being broken up by men with crowbars. Other workers were gathering the sulphur from the hissing mouths of the vents where it was condensing in dripping stalactites, ruby red in the hot centre and piercing yellow towards the margin.

My companions, shouting to one another above the incessant roar, disappeared into the swirling smoke to load their baskets. Astonishingly they seemed to be unaffected by the choking fumes. Soon they emerged, grinning cheerfully, with their baskets piled high with sulphur, and immediately, without pause for a rest, they set off briskly back down the slopes of the volcano towards their camp. I followed them, glad to escape into the purer air. The sky was cloudless and the atmosphere so clear that I could see beyond the green plains lying thousands of feet

Sulphur-gatherers

beneath us, across to the Java Sea on the horizon. To the east rose another group of mountains over which hung what I first took to be a cloud and then saw to be in fact a pall of volcanic smoke immeasurably greater than that of Walirang. I pointed to it and asked one of my companions its name. He shaded his eyes with his hand.

'Bromo,' he replied.

The distant smoke clouds had kindled my curiosity and when, that evening, Daan described Bromo as the most beautiful and famous of all the volcanoes of Java, Charles and I determined to visit it.

We left Tretes the next day in our jeep and drove eastwards along the coastal plain. From the road, Bromo appeared to be an unimpressive lumpish mountain, for it lies almost hidden among the wreckage of an even greater volcano. Thousands of years ago, this giant exploded and, like Krakatoa, blew away the greater part of its pyramid. Only its base remained as a mountainous ring enclosing a vast bowl nearly five miles across. But the energy of the volcano was not entirely spent in this tremendous eruption, for soon new vents opened inside the caldera, spewing ash and building fresh cones for themselves. None of these, however, has grown much higher than the encircling walls of the caldera and none, except Bromo itself, is still active. So it is that the approaching traveller on the plains sees not the erupting crater but only the uneven profile of the cliffs which surround it.

Evening was falling as we drove up the rocky track into a small village high on the outer slopes of the caldera. The mountains ahead were cloaked in cloud and the keeper of the inn at which we stayed told us that it was only in the early hours of the day that the volcano was free from mist. Accordingly we rose the next morning at half-past three. It was still dark and very cold. A little knot of villagers sat huddled in their sarongs beside a group of horses. Like the sulphur-gatherers, these hillmen were stocky and dark skinned, quite dissimilar from the lissom people of the plains. One man, who wore a luxuriant moustache, agreed to hire us two horses and to act as our guide to the crater.

After my experience on Walirang, I was surprised and delighted to find my mount to be an energetic creature, full of spirit. The man trotted barefoot behind us, occasionally whacking the horse's rump with a switch. I tried to dissuade him for my horse was quite frisky enough as it was, but I need not have concerned myself, for the horse good-humouredly ignored this treatment and only broke into a sudden gallop when the old man, running alongside it, bawled 'Whoosh' directly in its ear.

We reached the grass-covered lip of the caldera as dawn broke. Below us lay a desolate lunar landscape. The level floor of the bowl was partially veiled by skeins of wispy clouds. In its centre, almost a mile away, rose the peak of Batok, a stark symmetrical pyramid, its steep grey sides fluted with ravines and gullies. Bromo lay to the left, a low hump less shapely than Batok but more dramatic, for from its rounded crest issued an immense pillar of smoke. Beyond, dim in the early morning light, stretched the jagged wall which formed the caldera's further rim. We paused for a few minutes, awestruck. There was no sound except the continuous roar of Bromo.

With a 'Whoosh', the hillman urged the horses down the precipitous sandy track which descended to the caldera floor. The sun was rising, tinging the rolling volcanic smoke above us with a sullen pink. As the air warmed, the wisps of cloud dissolved, revealing a wide flat plain which stretched uninterrupted and featureless to the foot of Bromo and Batok. This 'sand-sea', as it was vividly but somewhat inaccurately named by the Dutch, is composed of grey volcanic dust which has been thrown out

Porters beside the extinct cone of Batok

from the vents and which, spread by the wind and rain, is slowly filling the caldera bowl. There are no lava flows around Bromo like those which treacle down the volcanoes of Hawaii, for the Javanese volcanoes are typified by a very viscous lava which solidifies at comparatively low temperatures. It is this characteristic which makes their eruptions so catastrophically violent, for as the molten lava in the deep furnaces of the earth's crust rises into the throats of the volcanoes, it cools slightly, solidifies and chokes the vent. The pressure beneath this plug then mounts until finally it becomes so great that the entire mountain explodes.

The horses trotted briskly across the barren plain until at last we reached the foot of Bromo. We left them and climbed up the steep muddy slopes towards the crater. At last we stood on the rim and gazed down into the cauldron. Huge volumes of smoke were pouring from a gaping hole in the bottom of the crater three hundred feet below, boiling out in gusts of such violence that the ground beneath our feet trembled. It thundered into the air in a creamy grey column, rising vertically, billowing and writhing, until as it approached our level the wind caught it and deflected it to one side so that its showers of hot grey dust fell to form a livid scar down the inner side of the crater.

We ventured fifty feet down the slopes of powdery ash which, as we kicked footholds, slithered downwards in tiny avalanches. The noise from the volcano was overwhelming, and the scale of the power unleashed beneath us frightening in its immensity. I looked back and saw the old man gesturing urgently to us to come back. Invisible pockets of heavy poisonous gases filled many of the hollows. If we unwittingly walked into one we might never return.

For centuries, the local people have made offerings to Bromo, lest it should burst into activity and annihilate the surrounding villages. In past times, it is said, human sacrifices were made, but today it is only coins, chickens and bolts of cloth which are cast into this hellish hole.

Such a ceremony took place only a few weeks after our visit.

Charles Lagus filming in the crater

We were told that crowds gathered on the brink of the crater to make their offerings. Bolder, less superstitious people climbed down the interior, as we had done, to snatch back the gifts from the maw of the god. During the scramble for one of the more valuable offerings, one of the men lost his footing and tumbled down the steep slopes. The assembled crowd watched him fall. No one made any attempt to rescue him and his body was left, like a broken toy, motionless in the depths of the crater.

A superstitious custom, which claims to pacify a spirit capable of bringing death to a whole population, is not easily discarded: perhaps the belief in the necessity of a human sacrifice has not yet entirely faded from the minds of the country people.

We returned to Surabaya the next day. There was no room in Daan's garage for our jeep, so we left her on the gravel drive outside our window. We had little fear that she would be stolen for we immobilized her by removing vital parts of her engine.

In the morning, we all climbed in to drive into the town. The engine started at a touch. Charles slid her into gear but the back wheels remained motionless. We investigated and found to our horror that a thief during the night had unscrewed the half-shafts and withdrawn them from the hubs with the result that the wheels no longer engaged with the main shaft. To us this seemed a disaster but Daan, though distressed, was not seriously perturbed and only mildly surprised.

'Dear me,' he said: 'There used to be a vogue for stealing windscreen wipers, but as everyone now locks them away and only fits them when it rains, perhaps that has gone out of fashion and the new craze is for half-shafts. Or maybe it is merely that someone has given an order for a pair at the thieves' market. I'll send the gardener down there early tomorrow morning. He'll probably find them, because it is usually considered only fair to give the rightful owners a chance to buy their things back.'

The half-shafts reappeared the next morning, but at a cost of several hundred rupiahs.

The day of our departure for Borneo had arrived. After the experience with the jeep's half-shafts we were nervous about the safety of our equipment, for we had twenty separate boxes, some of them containing cameras worth several hundred pounds. Transporting them from jeep to the customs shed, out into the dockyard, along the quay and into the security of a locked cabin seemed a formidable and hazardous operation. Daan was insistent that to leave any one piece unattended was tantamount to giving it away. So we decided that Daan would remain with the initial dump at the customs barrier and argue with the customs officer; I would go ahead to the ship, find our cabin and lock up the equipment as it arrived, while Charles would accompany each load with the porters whom we would be forced to employ. This seemed foolproof.

It began well. We fought our way through the jostling crowd to the customs shed and piled everything into a big heap on the barrier. Daan then opened negotiations and I, waving my ticket and passport, pushed my way into the dockyard. At last I found our ship, a large cargo boat, moored by one of the more distant quays, but I was disconcerted to see that the notice on the jetty proclaimed that she was not due to sail for another four hours, three hours later than we had been told in the shipping office. There were no gangways lowered and no officers in sight. The only way to get on board was along a narrow plank leading to a small black opening in the side of the ship which obviously led to the lower decks. Porters and seamen were swarming in and out. As I stood wondering what to do, debating whether to return and postpone the whole operation, I saw in the distance a trolley load of our equipment trundling down the jetty. If I were not ready to receive everything our carefully laid plan could fail. I joined the queue of porters and walked up the plank. Inside the ship it was dark, unbearably hot, and full of the smell from the half-naked begrimed porters, who crowded so tightly around me that I could hardly force my way through. I suddenly remembered that in the breast pocket of my shirt I was carrying all my money, my fountain pen, my passport and ticket. I clapped my hand over the pocket. It landed not on cloth but on someone else's hand. I gripped it as hard as I was able, slowly bent it back and removed my wallet from its fingers. Its owner, a sweating half-naked man with a dirty cloth tied round his forehead, glared at me savagely. The porter packed tightly round me murmured aggressively. I decided that in the circumstances it would be better to be gently reproving than to attempt an impersonation of an avenging fury, but the only word I could think of was 'Tidak. No'.

At this the pickpocket laughed a little nervously, which encouraged me, and with a thumping heart I thrust my way, in as dignified a manner as I could manage, up an iron ladder and on to the upper deck.

Charles stood on the jetty beneath, standing guard over the trolley and the baggage.

'Don't whatever you do come through the lower deck,' I called to him. 'I've just had my pocket picked.'

'I can't hear,' Charles called back, over the noise of the rattling cranes and yelling crowds.

'I've just had my pocket picked,' I roared.

'*Where* do you want the baggage?'

I gave up trying to tell him of my troubles and answered his question by pointing vigorously at the deck. At length he understood what I was proposing. I found a rope and threw it down to him and piece by piece I hauled the baggage up and dumped it on the deck. As the last box came over the rail Charles disappeared. Two minutes later he emerged on to the upper deck, looking rather ruffled and out of breath.

'You won't believe it,' he panted, 'but I've just had my pocket picked.'

It took us another hour before we and all our luggage were assembled in the safety of our cabin. The experience, however, had taught us a lesson. Never again did we pass through a crowd without one hand clamped on our wallets and the other clenched ready to defend ourselves. Indeed the habit became so ingrained that when, having been transferred by air in the space of three days from the markets of Jakarta to the rush-hour crowds in London, it was only by the narrowest margin that a stranger, who inadvertently jostled me at Piccadilly Circus, escaped a quite unwarranted punch on the jaw.

15

Arrival in Borneo

For four days our ship ploughed steadily northwards through the calm blue waters of the Java Sea towards the little town of Samarinda, which lies on the east coast of Borneo at the mouth of the Mahakam, one of the largest rivers on the island. We hoped that somehow we should be able to travel up this river into the country of the Dyaks to look for animals, and we had the names of two people who might be able to help us do so: Lo Beng Long, a Chinese merchant in Samarinda to whom Daan had written, and Sabran, a hunter and animal collector who lived a few miles upriver.

At dawn on the fifth day we steamed into the port. Lo Beng Long was on the quay to meet us and for the whole of that day he drove us round the town introducing us to all the officials whose approval we must have before we could leave for the interior.

Lo Beng Long had already reserved a motor launch for us. Her name was the *Kruwing*. She lay tied to the jetty, rocking gently in the foul garbage-strewn water of the harbour. She was forty feet long from stem to stern, powered by a diesel engine with a wheelhouse amidships and a cabin for'ard. She was roofed with tattered canvas, and staffed by a crew of five men, headed by an old cadaverous-faced captain who was referred to as 'Pa'.

Pa grudgingly admitted that he might be ready to sail the next day, so we hurriedly completed our last task of buying enough stores for us to live independently on the boat for a month. We returned from the market that evening laden with pots and pans, sacks of rice, bundles of pepper, cones of crystal-line palm sugar neatly wrapped in banana leaves, and a bag of small dried octopuses. This last purchase had been Charles', for he maintained that it would provide a stimulating dietary change after a week or so of rice. We had also bought sixty cakes of crude salt and several pounds of blue and red beads which we planned to use for barter with the Dyaks.

The first night we moored at Tenggarong, a line of sleazy wooden shacks stretching for a mile along the left bank. I had hoped we would steam throughout the night, for I was impatient to reach the Dyaks, but this Pa refused to do for he rightly said that there was a danger that in the darkness our boat might run afoul of the floating logs which littered the river, or might collide with other craft, few of which carried navigation lights.

A crowd gathered on the jetty to stare at us and Daan went ashore to gossip. This was the village in which we had been told that Sabran the hunter lived, but Daan could find no one who had heard of him. The crowd stayed to watch us eat, but as soon as that entertainment was concluded and darkness fell, they dispersed.

There was room for three of us to sleep in the for'ard cabin, but on this, our first night aboard, the bunks were still cluttered with unstowed baggage. I therefore decided to take the camp-bed off the boat and sleep in airy isolation on the jetty. It was pleas-antly cool after the heat of the day, and I quickly fell asleep. Barely had I done so than I woke up with a start. Within a few feet of my face, a large bewhiskered rat crouched gnawing a palm nut. Behind it, ghostly in the moonlight, several others were foraging among the trash on the jetty. Another trailed its long scaly tail around the bollard to which our mooring rope was attached. I earnestly hoped that we had left nothing lying on the deck of the *Kruwing* which would attract these loathsome creatures on board. I watched them for a long time as they scuffled around me but I felt disinclined to do anything. The thought of putting my bare feet among them filled me with revulsion, and while I remained inside the tent of my mosquito net I illogically felt quite secure.

At last I went to sleep again but it seemed that no sooner had I closed my eyes than I was woken once again, this time by a voice calling 'Tuan, tuan' in my ear. A young man stood by me with a bicycle. I looked at my watch; it was not yet five o'clock.

'Sabran,' said the man, pointing to his chest.

I swung my feet out of bed, hitched my sarong around my naked body, and did my best to sound a little more welcoming than I felt. I called to Daan who eventually stuck his tousled head through the cabin hatch. The young man explained that word had reached him last night of strangers who had been asking for him. Anxious not to miss us, he had bicycled from his house several miles away to be at the jetty before dawn broke and we departed. This eagerness and zeal, as we were to discover later, was wholly typical of Sabran. In his early twenties and naturally enterprising, he had several years before worked his passage on a merchant ship to Surabaya to see the big city about which he had heard so much in Samarinda. He took a well-paid job for a short time, but the poverty and squalor of Surabaya had so horrified him that he had decided to return and work for less money in his native forests. Now he lived in Tenggarong supporting his two sisters and his mother by undertaking commissions to catch animals. It was clear that his help would be of great value to us and we suggested he might join us. Sabran agreed immediately. He pedalled away and by the time we were finishing our breakfast he had returned carrying all his gear in one small fibre suitcase. Before we realized what had happened he was washing our dirty breakfast dishes in the stern. Sabran, it was clear, was going to be a real asset.

After breakfast we sat down with Sabran to discuss our plans. I drew pictures of the animals in which we were interested and he told us their local names and where they were to be found. We were particularly anxious to see the proboscis monkey, a

Sabran

spectacular creature which lives only in the coastal swamps of
Borneo. It was easy to draw for, alone among monkeys, the male
possesses an enormous pendulous nose. Sabran immediately
recognized my clumsy sketch and said that he could guide us
to a place a few miles up the river where we could see them.

We reached it that evening. Pa cut the engine and we drifted
slowly with the current close to the tall forest which lined the
banks. Sabran sat in the bows shading his eyes with his hands.
At last he pointed excitedly to a place on the bank a hundred
yards ahead. The monkeys were sitting in the thick vegetation
at the water's edge nonchalantly pulling off leaves and flowers
and cramming them into their mouths. There were about twenty
in the group. They looked at us solemnly and without fear. The
majority were youngsters and females, uniformly red in colour.
They looked absurdly comic for their long noses were snubbed
and turned up like a circus clown's. The old male who governed

the troop, however, had an even more ridiculous appearance. He sat high in the crotch of a tree dangling his long tail like a bell rope. His red coat ended sharply at his waist, his pelvis was covered in white fur as was his tail, his legs were dirty grey, so that he appeared to be wearing a red sweater and white bathing trunks. His most astonishing characteristic, however, was his vast flaccid nose which hung down his face like a red squashed banana. It was so large that it seemed a positive encumbrance to him for it got in the way as he ate and he was forced to make a detour with his hand around and under his nose to get his food into his mouth. We drifted closer and closer until at last the monkeys took fright and, leaping away with an agility surprising for such big creatures, they disappeared into the forest.

The diet of these extraordinary monkeys is exclusively vegetable. None had ever survived for any length of time outside the tropics because no one had then devised an adequate substitute for the particular leaves on which the proboscis monkeys feed. We did not, therefore, make any attempt to catch one, but for several days we cruised up and down the banks filming them.

Every morning and evening they came to the riverside to feed, but during the hot hours they slept out of sight in the shade of the forest. In the daytime, therefore, we looked for other creatures and in particular for the man-eating crocodiles, which, so we had been told in Samarinda, infested the river in vast numbers. To our disappointment we found not the slightest sign of them. There were, however, many beautiful birds in the forest, and hornbills in particular were especially abundant. We were very interested in these creatures for their breeding habits are among the most extraordinary in the bird kingdom. They nest in holes in trees and the hen, when she settles down to incubate her eggs, is imprisoned in the hole by the cock who walls up the nest entrance with mud. He leaves a small window in the centre and through it he passes food to the hen. She keeps her prison scrupulously clean, casting out the droppings every day, and she remains in the nest until her young are hatched and fledged. When they are ready to fly she breaks down the wall and the whole family leaves the nest.

Soon we left behind the semi-cultivated country and entered an area where the banks were still clad in high forest. On the fifth day we came to a small village. The river was low and the edge of the water was separated from the palm trees on the bank above by a hundred yards of brown sticky mud. We moored by a small stage of floating tree trunks and went ashore, crossing the mud along a line of notched logs laid end to end.

At the top of the bank, surrounded by palm trees and bamboo, stood the first Dyak village we had seen. It consisted of a single wooden dwelling a hundred and fifty yards long, roofed with wooden shingles and standing ten feet above the ground on a forest of stilts. The veranda, which ran along the front, was crowded with villagers watching our arrival. We climbed up the steep notched pole leading into the house where we were received by a dignified old man, the *petinggi*, or headman. Daan greeted him in Malay, we introduced ourselves and the *petinggi* led us down the longhouse to a place where we might sit and smoke together and tell him of our plans.

We walked along a nave of immense ironwood columns which ran the length of the building. Many of them were decorated near the roof with writhing wooden beasts. Above these carvings were tied sheafs of bamboo sticks, cleft at one end to hold eggs and rice cakes as gifts to the spirits. From the rafters hung dusty buckled trays on which offerings had been made, and bundles of withered leaves which had cracked and parted to reveal the yellowing teeth of human skulls.

Between the pillars, in special racks, lay long drums. The corridor was floored with gigantic adze-hewn boards and flanked on one side by the veranda and on the other by a wooden wall beyond which lay private rooms each occupied by a family.

This magnificent building, however, was in decline. In places, the roof had collapsed and the rooms beneath had been abandoned. The parchment covering the heads of many of the drums was cracked and some of the floor boards were rotting and riddled with termite holes. At either end of the longhouse, wooden pillars stood supporting nothing among the sprouting bananas and bamboos, showing that the house in its prime had been of even greater length. Most of the villagers who sat on the veranda watching us as we passed had discarded their traditional Dyak costume and wore singlets and shorts. But though they had been able to adopt these new fashions from the outside world, the older people could not shed all the signs of the ceremonies which they had undergone in their youth when their ancient customs were still unsullied. Most of the women had pierced earlobes which, having been weighed since their youth with heavy silver rings, were now so stretched that they hung down to their shoulders. Their hands and feet were blue with tattooed decorations. Both men and women chewed betel nut which had coloured the inside of their mouths brown red and had eroded their teeth into decayed black stumps. Red star-shaped stains spangled the corridor floor, showing where they had spat the saliva which is so abundantly stimulated by the nut. Few of the young people chewed it; instead, they had plated their teeth with gold in the best Samarinda manner.

The *petinggi* told us that Roman Catholic missionaries had settled a little way from the longhouse and had built a church and a school. As his villagers adopted the new faith, they threw away the human skulls which had been won in war and which they had venerated for generations, and pasted lithographed religious prints on their walls. But though the missionaries had been working for over twenty years here, they had so far converted less than half of the villagers.

That evening, as we sat eating supper on the deck of the *Kruwing*, a Dyak came stepping nimbly down the logs across the mud, carrying in his hand by its trussed legs a flapping white chicken. He climbed on board and presented us with the bird.

'From *petinggi*,' he said gravely.

He also brought a message. There would be music and dancing

in the longhouse that night in celebration of a wedding. If we wished to come we would be welcome.

We thanked him, gave him some cakes of salt as a return present for the *petinggi*, and told him that we gladly accepted the invitation.

The villagers were gathered in a wide circle on the veranda. The bride, a beautiful girl with sleek black hair drawn back from her oval face, sat with lowered eyes between her father and her husband. She was dressed splendidly in a scarlet beaded headdress and a richly embroidered skirt. In front of her, by the flickering light of coconut-oil lamps, two older women gyrated in a stately dance. On their heads they wore small beaded caps hung with fringes of tigers' teeth and they held in their hands sprays of the long black and white tail feathers of the rhinoceros hornbill. On the opposite side of the circle, a man, naked to the waist, sat playing an endless repetitious tune on six gongs lying in a rack. The *petinggi* rose as we entered and sat us in places of honour by his side. He was very curious about the green box I had brought with me and I tried to explain that it was able to capture sounds. He was mystified. Surreptitiously I set up the microphone, recorded a few minutes of the gong music and, during a pause in the music, I played it back to him on a small speaker.

The *petinggi* got to his feet and stopped the dance. He called the gong player to bring his instrument to the centre of the circle, gave him instructions to play a new and more lively tune, and then invited me to record it. The children, excited by this new turn of events, chattered so loudly that I could hardly hear the music, and, fearful that I should disappoint him with an inferior recording, I sat with my finger to my lips in an attempt to keep them quiet.

The playback was a vast success and when it was finished the longhouse echoed with gales of laughter. The *petinggi* regarded this performance as a personal triumph and, assuming the role of impresario, he began to organize the numerous volunteers who wished to sing into the microphone. While he was so

Recording in the Dyak longhouse

engaged I glimpsed the bride sitting lonely and ignored and I became suddenly ashamed of having disrupted the festivities in her honour.

'No,' I said. 'Machine now tired. Not work again.'

Within a few minutes the dances were resumed but in a half-hearted way. Everyone's eyes were fixed on the machine by my side waiting for further miracles. I picked it up and took it back to the ship.

The next morning the festivities continued. Now it was the turn of the men. They danced to the music of drums and a *gambus*, a three-stringed guitar-like instrument. Most of them wore only the long traditional loincloths and carrying shields and swords they pranced in slow motion in front of the long-house, occasionally leaping in the air with wild cries. Among them moved an even more impressive figure, cloaked from head to foot in palm leaves and wearing a long-nosed wooden mask

196

painted white, with flaring nostrils, long fangs and two circular mirrors for eyes.

My worries that we had inexcusably invaded the villagers' privacy were perhaps misplaced for the Dyaks themselves were as curious and prying about our doings as we were about theirs. Every evening they came down to the ship and sat on board watching our strange method of eating with knives and forks and staring fascinated at our equipment. Stripping the inside of the recorder was the highlight of one evening, and a certain success was a demonstration of flashlight photography.

In turn we ourselves became emboldened to wander into the private rooms in the back of the longhouse. A few of them contained low beds hung with filthy torn mosquito nets, but most of the people ate, sat and slept on mats of split rattan on the floor. This lack of furniture seemed to be no loss to anyone except to the small babies, and the Dyak mothers solved this problem by binding the infants in long loops of cloth and hanging them from the ceiling. The children slept in this upright position and when they cried their mothers merely pushed them so that they pendulumed gently to and fro.

We told everyone we met that we would give generous rewards for any animals that they brought to us for we knew that even the least skilful of the Dyak hunters would catch more animals in a week than we could in a month. Unfortunately, no one seemed interested in our proposition. In one of the rooms of the longhouse, I saw lying on the floor a pile of long wing feathers of one of the most beautiful and spectacular of Borneo's birds, the argus pheasant.

'Where is this bird?' I said in anguish.

'Here,' replied the woman of the household, and she pointed to sections of the bird's carcass lying plucked and jointed in a calabash ready to be cooked.

I groaned. 'But I can give many, many beads for such bird.'

'We hungry,' she replied simply.

It was clear that no one believed we would give rewards large enough to make it worth their while to bring live animals to us.

One day, as Charles and I returned from filming in the forest, we met one of the older men who was one of our regular visitors on the *Kruwing*.

'*Selamat siang*,' I said. 'Peace on the day. Can you catch animals for me?'

The old man shook his head and smiled.

'Look well at this,' I said and took out of my pocket an object I had picked up in the forest. It appeared to be a large polished marble, striped with orange and black. Suddenly it unrolled and revealed itself as a remarkably handsome giant millipede. It trundled steadily on its numerous legs across the palm of my hand, cautiously waving its black knobbly antennae.

'We want many different animals; big ones, small ones; if you bring me this,' I said, pointing to the millipede, 'I give you one stick of tobacco.'

The old man goggled.

This was indeed an extravagant price to offer for such a crea-ture but I was anxious to emphasize how keen we were to obtain animals of all sorts. We left the old man standing astounded and I felt pleased with myself.

'If we are lucky,' I said to Charles, 'he may be the first recruit to our animal-catching team.'

The next morning I was wakened by Sabran.

'Man come with many, many animals,' he said.

I jumped enthusiastically out of bed and rushed on deck. It was the old man. In his hands he was holding a large gourd, as though it were an object of inestimable value.

'What?' I asked eagerly.

In reply he carefully emptied its contents on to the deck. At a rough estimate, there were between two and three hundred small brown millipedes, hardly dissimilar from the ones I could

Feeding Benjamin the bear

His skin was covered in small scabs each of which concealed a white wriggling grub and after every meal we had to clean and disinfect these tiny wounds.

It was not until several weeks later that he began to walk but as soon as he did so his character appeared to change. As he tottered and swayed across the ground, smelling everything and grumbling to himself, he seemed no longer to be an impatient demanding creature, but rather an endearing puppy, and we both developed a strong affection for him. When at last we brought the collection back to London, Benjamin was still needing milk from a bottle, and Charles decided that instead of handing him over to the Zoo with the other animals he would keep the bear for a little longer in his flat.

Benjamin was now four times as large as when we had first found him and had developed large white teeth with which he was well able to defend himself. Although most of the time he was well mannered and peaceable, sometimes, if he was baulked in one of his investigations or games, he would fly into a furious temper, slashing wildly with his claws and growling angrily. In spite of ripped linoleum, chewed carpets, and scratched furniture, Charles kept him until he had learned the knack of lapping milk from a saucer and was no longer dependent upon a bottle. Only then did Benjamin go to the Zoo.

16

Charlie, the Orangutan

Of all the animals of Borneo, the creature I was most anxious to find was the orangutan. This magnificent ape, whose name translated from the Malay means 'man of the forest', is found only in Borneo and Sumatra, and even in these territories it is restricted to comparatively small areas. In northern Borneo it is already becoming very scarce and although everyone we had met here in the south of the island claimed that the animal was still abundant, very few people seemed actually to have seen one. We decided to devote our last days on the Mahakam to an intensive search and to travel slowly back downstream, calling not merely at the bigger villages but at every small hut and landing until we discovered someone who had caught sight of the apes recently.

Luckily, we did not have to travel far. On the first day of our return journey we stopped at a small shack built on a floating ironwood pontoon tied to the bank. The owner lived by trading with the Chinese boats which came up the river from Samarinda, exchanging for their goods the crocodile skins and rattan brought to him by the Dyaks from the forest. Several such Dyaks were standing on the landing-stage as we came alongside. They were wild-looking men, their straight black hair cut in a fringe across their foreheads, naked except for their loincloths, and carrying long *parangs* in tasselled wooden sheaths. They told us that within the past few days, families of orangutans had been raiding the banana plantations near their longhouse. This was the news we had been seeking.

'How far is your village?' Daan asked.

One of the Dyaks looked at us critically.

'Two hour for Dyak,' he said, 'four hour for white man.'

We decided we would go, and the tribesmen agreed to guide us and help carry our baggage. Rapidly we unloaded our equipment, a few spare clothes and a little food, all of which the Dyaks stowed neatly in their rattan carriers. Leaving Sabran on board the *Kruwing* to take care of Benjamin and the rest of the animals, we followed the men up on to the riverbank and into the forest.

We soon discovered why the Dyaks had thought it unlikely that we would make the journey as fast as they, for the path, such as it was, ran through very marshy forest, and across a series of swamps. We waded through the shallower pools, but when we came to deeper ones, we crossed by balancing along slender slippery tree trunks which often lay a foot beneath the surface of the muddy water. The Dyaks scarcely slackened their pace as they came to these obstacles and walked across them as though they were going down a main highroad, but we had to take a great deal more care over our equilibrium, feeling cautiously with our feet for the invisible log, knowing that if we stumbled, we should fall into deep water.

It took us three hours to reach the longhouse. It was more ramshackle than the one we had visited earlier: the floor was not made of boards but of thin strips of split bamboo and the interior had no private rooms, being only roughly divided by a few flimsy screens. Our guides led us through the crowded house and showed us to a corner where we might put our things and eventually sleep. It was already evening, and we cooked our supper of rice over a small fire burning nearby on a hearth of stones. By the time we had finished, it was dark. We folded up our bush jackets, put them beneath our heads and lay down to sleep.

Normally, I am able to sleep moderately well on hard boards, but I do require a certain degree of quietness and the longhouse was filled with noise. Dogs prowled everywhere unchecked, yelping as someone kicked them out of the way. Fighting cocks clucked and crowed from cages tied to the wooden walls. Not far away from us, a group of men sat playing a gambling game, spinning a top on a tin plate, clapping half a coconut over it and then loudly calling bets. Quite close to me, a circle of women were chanting around a curious rectangular edifice, shrouded by cloth curtains suspended from the rafters. A few of the villagers, undisturbed by this clamour, lay about in untidy groups asleep – some outstretched, some sitting with their backs against the walls, some squatting, knees up, with their heads resting on their forearms.

In an endeavour to shut out the din, I draped a spare shirt over my head. This muffled some of the sounds but concentrated my attention on what I could hear through my pillow. Directly below me, several smelly pigs squealed and grunted as they rooted in the refuse which had been tipped among the stilts of the house. The resilient bamboo floor produced a swishing creaking noise as the villagers pattered about on it. Each time someone moved near me, my body bounced slightly, and a person walking twenty yards away made the floor squeak so loudly that it seemed he was jumping over my head. I was well able to make this assessment for only too often people did in fact step over my prostrate body.

Fortunately the yapping, clucking, chattering, shouting, chanting, grunting and squeaking combined into a noise so constant and unvarying that eventually it became monotonous and I went to sleep.

I woke in the morning, stiff and barely refreshed and, together with Charles and Daan, went down to bathe in a small river which flowed a hundred yards away from the house. It was already full of naked splashing people, the men washing together in a deep pool and the women in another a few yards downstream. We sat in the warm sun on a neat wooden platform and dangled our legs in the sparkling stream. Our guide was also washing and after we had finished we walked back with him to the longhouse.

On the way we passed a newly built shelter thatched with atap. I noticed lying on a platform beneath it a long pillar of brown wood which had been carved at one end in the shape of a human figure. Close by a huge water buffalo was tethered.

'That,' I asked, indicating the pillar, 'what is it for?'

'Man dead in longhouse,' was the reply.

'Where in longhouse?'

'Come,' he replied, and we followed him up the step-pole into the house.

'Here,' he said, pointing to the draped platform around which the women had been chanting on the previous night. I had been sleeping unsuspectingly within a few yards of a corpse.

'When did the man die?' I asked.

Our guide thought for a moment. 'Two year,' he said.

He told us that a Dyak funeral is a very important event. The richer a man has been during his lifetime, the more elaborate and lengthy the funeral feast that his children must provide in his honour after his death. The dead man in the longhouse had been a person of consequence but his children were poor, and it had taken them two years to save up enough money to be able to provide a feast that would be worthy of his memory. During that time the body had been placed high in a tree, exposed to the sun and the wind and the depredations of insects and carrion-feeding birds.

Now the time for the feast had arrived and the bones had been brought down from the tree to lie in state before their final interment.

That afternoon the village musicians carried gongs down from the longhouse and played for half an hour, while a group of mourners danced round the pillar which they had erected in the clearing. It was a short and unimpressive ceremony.

'Finish?' I asked my friend.

'No. We kill buffalo when ceremony end.'

'When you do that?'

'Maybe twenty, thirty days' time.'

The celebrations would continue every day and night for the next month with increasing frequency and duration. On the last day, during a final orgy of dancing and drinking, all the villagers would descend from the longhouse with *parangs* in their hands and circle the buffalo until at the climax of the dance they would close in upon it and hack it to death.

We had said that we would give rewards to anyone who was able to show us a wild orangutan, and the first claimant woke us at five o'clock the next morning. Charles and I snatched up cameras and followed the man at a slow trot into the forest. When we arrived at the place where he had seen the animal we found freshly chewed rinds of durian fruit, the orangutan's favourite food, scattered over the forest floor. In the trees above, we saw a large platform of broken branches on which the ape had slept during the night. But though we searched for an hour we could not find the animal itself, and we returned disappointed to the village.

That morning we made three more unsuccessful sallies into the forest and the next day another four, so eager were the villagers to earn the reward of salt and tobacco. On the third morning, once again a hunter came in to say that he had just seen an ape and once more we scampered after him, squelching through deep mud, oblivious of the savage thorns which snatched at our sleeves, anxious only to reach the spot before the ape moved away. Our guide trotted ahead over a fallen tree trunk which bridged a deep creek. I followed him as fast as I could with the heavy camera tripod on my shoulder. As I crossed I held on to a branch to balance myself. It snapped. With my other hand gripping the tripod I was unable to regain my balance, my feet slipped and I fell into the river six feet below, striking my chest heavily on the trunk as I dropped. I struggled to my feet in the water, winded and with an agonizing pain in my right side. Before I could reach the bank the Dyak was beside me.

'*Aduh*, tuan, *aduh!*' he murmured compassionately, clasping me to him with heart-warming sympathy. With no breath in my lungs I could do nothing but groan feebly. He helped me out of the water and up the bank. The blow had been so severe that the binoculars, which I had been carrying under my right armpit, were smashed in two. I felt gently over my chest, and from the swellings and the great pain I was sure I had cracked two of my ribs.

After I had recovered my breath, we walked slowly onwards. After a while the Dyak began to imitate the orangutan's call, a combination of grunts and ferocious squeals. Soon we heard an answer. We looked up and saw, swaying in the branches, a huge hairy red form. Rapidly, Charles set up his equipment and started to film while I rested on a tree stump nursing my aching side. The orangutan hung above us, baring his yellow teeth and squealing angrily. He must have been nearly four feet tall and weighed perhaps ten stone – I was sure that he was larger than any I had ever seen in captivity. He climbed to the top of a slender branch until it bent beneath his weight and curved downwards towards a neighbouring tree. Then he stretched out one of his long arms and lumbered across. Occasionally he broke off small branches and threw them down at us in fury, but he seemed to be in no hurry to escape. Before long we were joined by other villagers, who helped us to carry our gear as we followed the animal and enthusiastically cut down saplings to give us a clear view of him. We had to pause every few minutes for the damp forest in which we were working abounded with leeches. If we stayed in one particular place for long they came looping across the leaves of the undergrowth like small thin worms. When they reached us, they crawled on to our legs and dug their heads into our flesh, sucking blood until they were swollen to many times their original size. Preoccupied with watching

the ape, we often did not notice them until the Dyaks thought-
fully pointed them out and shaved them off with their knives,
so that the places in which we had filmed were marked not
only by the fallen saplings but by the severed oozing bodies of
the leeches.

At last we decided we had secured all the film we needed
and began to pack up.

'Finish?' asked one of the Dyaks.

We nodded. Almost immediately there was a deafening explosion behind me and I turned to see one of the men with a
smoking gun to his shoulder. The ape had not been badly hit
for we heard it crashing away in the distance to safety, but I was
so angry that for a moment I was speechless.

'Why? Why?' I said in fury, for to shoot such a human creature seemed to amount almost to murder.

The Dyak was dumbfounded.

'But he no good! He eat my banana and steal my rice. I
shoot.'

There was nothing I could say. It was the Dyaks who had to
wrest their livelihood from the forest, not I.

That night as I lay on the floor of the longhouse, my ribs
stabbed pain every time I took a breath and my head began to
ache. Suddenly a chilling tremor shot through me and I started
to shake uncontrollably, my teeth chattering with such violence
that I could hardly speak intelligibly. I had malaria. Charles dosed
me with aspirin and quinine and I spent a bad night, further
disturbed by the wailing and beating gongs of the continuing
funeral ceremony. When I woke in the morning all my clothes
were soaked with sweat and I felt miserably ill.

By midday I was sufficiently revived to contemplate the
journey back to the boat. We had achieved our purpose in
coming to the village – we had filmed the orangutan – and
we had to return. We took the journey very slowly and I
rested many times on the way, but nonetheless I was very
relieved when at last we reached the *Kruwing* and I was able

to sweat out the rest of my fever in the comparative comfort
of a bunk.

When we had first joined the *Kruwing*, the crew had been
somewhat reserved: no one had spoken to us except Pa and he
I had scandalized on the first evening by suggesting that we
should travel throughout the night. I had felt that they regarded
us as ignorant though harmless lunatics.

As the weeks passed, however, their attitude had changed and
now they were genuinely friendly. Pa was full of helpful suggestions: if he saw a movement in the forest ahead, he would of
his own accord ring down half-speed to the engine room, and
call to ask us if we wished to film the animal he had spotted.
The *masinis*, the engineer, a large burly man who never wore
anything but the blue overalls of his calling, was unashamedly a
townee. The jungle held no fascination for him and he was so
little interested in the Dyak longhouses that he seldom bothered
to go ashore but sat on the deck above his engine room, mournfully plucking bristles from his chin with a pair of nail clippers.
He had a standard witticism which he produced at each settlement. '*Tidak baik*,' he would say. '*Bioskop tidak ada.*' No good,
there is no cinema.'

Joking indeed was one of our main pastimes during the long
hours of steaming down the wide river. It was a laborious process,
for the preparation of a new joke took several hours. Having
devised it, I required perhaps a quarter of an hour working with
my dictionary to translate it. Then I would go aft to the stern,
where the crew would be sitting brewing their coffee, and
painstakingly I would deliver my ham-fisted jest. Usually it would
be met with blank stares and I would have to return to work
out an alternative wording. Often I had to make three or four
attempts before I succeeded in conveying it, but when I did the
crew always laughed uproariously, more for my benefit, I suspect,
than at the joke itself. Once cracked, however, the jest was not

discarded but entered everyone's repertoire, to be resuscitated time and again during the next few days.

Hidup, the second engineer, we seldom saw for the *masinis* kept him in the engine room most of the day. One evening, however, he appeared on deck having shaved off all his hair. While he sat blushing, stroking his naked scalp and laughing at his own embarrassment, the *masinis* explained to us in detail that Hidup had had nits in his hair.

The deckhand, Dullah, was a wrinkled old man who spent a great deal of his time giving us lessons in Malay. He considered, with the best European educationists, that the best way to teach a language is never to allow the pupil to speak his native tongue. He would therefore come and sit by us and talk patiently and slowly, with exaggerated articulation, about any subject that came to his mind – the nomenclature of Indonesian costume, the varying qualities of rice – but always at such length and in such detail that after a few minutes we became hopelessly confused and were reduced to nodding knowingly and saying *'Ja, ja!'*

The fifth member of the crew, Manap the bos'n, was perhaps the most helpful of all. He was a young man, handsome and usually undemonstrative, but if he was at the wheel when we saw an animal in the forest we could rely on him to take us closer to the bank and to negotiate the hazardous shallows with more skill and daring than anyone else.

The most energetic person on board, however, was Sabran. He undertook the major share of cleaning and feeding the menagerie, he cooked most of our meals and if we left dirty clothes within his reach he washed them unasked. One evening I told him that when we left Borneo, we planned to travel eastward to Komodo to look for the giant lizards. His eyes glowed with excitement, and when I asked him if he would like to come with us, he seized my hand and pumped it up and down saying delightedly, 'Is OK, tuan, is OK.'

One morning Sabran suggested we should stop, for close by lived a Dyak friend of his named Darmo who in the past had helped Sabran to catch animals. It might be that he had trapped some recently which he would trade to us.

Darmo's home was a small stilted hut, squalid and dirty, and Darmo himself, an old man with long greasy hair hanging down his back and falling in an untidy fringe over his forehead, sat on the platform outside it, whittling a piece of wood. Sabran called up to him as we approached and asked if he had any animals. Darmo looked up and said in an expressionless voice, *'Ja, orangutan ada.'*

I scampered up the pole three steps at a time. Darmo pointed to a wooden crate, clumsily barred with strips of bamboo. Inside squatted a young very frightened orangutan. Cautiously I poked my finger inside to scratch its back, but the ape swung round with a squeal and tried to bite me. Darmo told us that he had caught the creature only a few days previously when it was raiding his plantations. In the struggle he had been bitten badly on his hand, and the ape had grazed its knees and wrists.

Sabran began negotiations on our behalf and Darmo eventually agreed to exchange the creature for all our remaining salt and tobacco.

Our first task on getting the ape back to the *Kruwing* was to transfer him from his original cage to a bigger and better one. This we did by placing the two face to face, drawing out the bars of the old cage, lifting the door of the new one and then enticing the orangutan into its new home with a bunch of bananas.

The little creature was male, about two years old, and we called him Charlie. For the first two days that we had him we left him alone, so that he might settle down in his new cage. On the third day, I opened his door and cautiously put my hand inside. At first Charlie snatched at my fingers and bared his yellow teeth trying to bite me. I persevered and at last he allowed me to bring my hand slowly towards him and scratch his ears and his fat paunch. I rewarded him with some sweet condensed milk. I repeated the process in the afternoon and Charlie behaved so well that I boldly offered him some condensed milk on the end of my forefinger. Charlie tentatively pursed his wide mobile lips and noisily sucked off the sticky milk without making the smallest attempt to bite me.

Most of that day I sat by his cage, talking softly to him and gently scratching his back through the wire cage-front. In the evening I had won his confidence sufficiently for him to allow me to inspect the wounds on his arms and legs. Gently I took his hand and stretched out his arm. Charlie watched me gravely as I took some antiseptic cream and spread it liberally on the graze on his wrist. The ointment looked remarkably like condensed milk and no sooner had I finished than Charlie promptly licked it off, but I hoped that enough remained to be in some degree effective.

We were all astonished at the speed with which Charlie settled down. Soon he was not only tolerant of my fondling but actively sought it. If I passed his cage without stopping to talk to him he would call sharply to me. Often as I stood tending the birds which chattered in the cage beside his, a long scrawny arm would slide out beneath the bars of his cage and tug at my trousers. So persistent was he that I was usually compelled to feed the birds with one hand and clasp Charlie's black gnarled fingers with the other.

I was anxious to let him out of his cage as soon as possible so that he should get some exercise, and for the whole of one morning I left his door open. Charlie, however, refused to come out. He seemed to regard his box not so much as a prison but as a house which he knew and preferred to the bewildering unknown world of the ship's deck, and he sat inside with an expression of brooding solemnity on his dark brown face, blinking his yellow eyelids.

I decided to try and lure him out with a tin full of warm sweet tea, a drink of which he had become very fond. As he saw it, he sat up expectantly, but when, instead of giving it him immediately, I held it outside the open door of his cage, he squeaked with irritation. He advanced to the door and peered out cautiously. I kept the tin beyond his reach until at last he had come right outside and holding on to the door he leant over to sip it. As soon as it was finished he swung himself back into his cage.

Charlie drinking tea

The next day, I opened the door and he came out of his own accord. For a short time he sat on top of his box while I played with him. I tickled his armpits and he lay back, baring his teeth in an ecstasy of silent laughter. After a few minutes, he tired of this and swung himself down to the deck. First he inspected all the animals, pensively poking his fingers through the wire. He lifted the cloth from Benjamin's box; the little bear cub brayed lustily, thinking food might be coming, and Charlie hastily retreated. He moved on to the hanging parakeets and managed to steal some of their rice with his crooked forefinger before I was able to stop him. He then turned his attention to the miscellaneous collection of objects which lay about on the deck, picking up each one, lifting it above his mouth and pressing it to his little squat nose in order to sniff it and assess its edibility.

I decided that the time had come for him to go back to his cage, but Charlie did not wish to go and slowly ambled away from me. My ribs were still so swollen and painful that I too could only move at his speed and the *masinis* was greatly amused at the spectacle of me chasing Charlie in slow motion and trying to sound commanding as I instructed him to return to his box. I succeeded only by bribery. I showed him an egg and then placed it at the back of his cage. With dignity Charlie clambered inside, bit the top off the eggshell and neatly sucked it dry.

From that day onwards it became part of the ship's routine that during the afternoon Charlie should have his ramble. The crew became very fond of him, but always treated him with circumspection. If he began to misbehave himself, they dared not be firm with him but called to us for assistance. So it was that when at last we sailed into Samarinda, Charlie was sitting at Pa's elbow in the wheelhouse, for all the world like an extra member of the crew.

Our trip in Borneo was over. We had berths booked in a large merchantman, the *Karaton*, which was sailing the next day for Surabaya. Charles, Sabran and I began to plan how we should transport all the animals and baggage on to the ship. We knew that it would have been insulting to ask the *Kruwing* men to demean themselves by acting as porters. We were therefore most moved when in the evening Manap came for'ard and said gruffly that Pa had asked the harbour authorities for permission to take the *Kruwing* alongside the *Karaton*, and that if we wished it, he and his companions would shift our equipment for us.

They worked with a will, hauling everything up the steep iron walls of the *Karaton*'s side, shouting cheerful farewells to the

Charlie enjoying an afternoon out on the boat

animals as they swung upwards in their cages. At last everything was on board, the animals under Sabran's care stowed in a quiet corner of the boat deck, and all our luggage safely locked in our cabins. As the final piece arrived, the entire crew, Pa, Hidup, *masinis*, Dullah and Manap lined up outside our cabin to say goodbye. One by one they shook us warmly by the hand and wished us *selamat djalan*. We were very sorry to leave them.

17

A Perilous Journey

The problem of getting to Komodo was not one which anybody in Surabaya seemed to be in the habit of solving. The island lay five hundred miles away, the fifth in the string of islands which stretches eastwards a thousand miles from Java towards New Guinea. None of the government officials we knew could tell us how to get there, so we began our own investigations.

The clerk in the shipping office had never heard the name before and we had to point it out to him on his map, printed above a tiny spot between the two large islands of Sumbawa and Flores. Most of the interlacing black lines curving across the map, representing the routes of his company's ships, seemed purposely to avoid it. One route only, looping eastwards along the chain of islands, seemed to offer any hope. It dipped down to a port in Sumbawa, swung up, and ran past Komodo before curving down again to Flores. Both of its ports-of-call in Sumbawa and Flores were within a reasonable distance of Komodo.

'This boat,' I said, pointing to the black line. 'When does this one sail?'

'Next boat, tuan,' the clerk replied cheerfully, 'two months' time.'

'In two months,' replied Charles, 'we are supposed to have been in England for three weeks.'

'Is there,' I asked, disregarding this pessimism, 'a small boat in Surabaya which we could charter to take us directly to Komodo?'

'There is not,' said the clerk. 'And if there were, you could not. Chartering boats makes much, much trouble. Police; customs; military; they will not give permits.'

The airline officials were a little more helpful. We discovered with their help that by flying north to Macassar, in the island of Sulawesi, we could catch a small plane which flew fortnightly to Timor and landed on the way at a place in Flores called Maumere. Flores is a banana-shaped island, two hundred miles long. Maumere is within forty miles of the eastern end; Komodo lies five miles beyond its western end. The map also showed a road running along the length of the island. If we could be sure of being able to hire a car or a lorry in Maumere to take us along this road, our troubles were solved.

We found several people in Surabaya who had heard of Maumere, but none who had been there. Our most authoritative informant was a Chinese man who had a distant relative running a store in Maumere.

'Autos?' I asked him. 'Are there many autos?'

'Many, many, I am sure. Please, allow me to send a telegram to my relative, That Sen. He will arrange everything.'

We thanked him with very real gratitude.

'It's easy,' I said to Daan that evening. 'We fly to Macassar, connect with a plane to Maumere, discover our Chinese friend's brother-in-law, hire a lorry, drive two hundred miles to the other end of Flores, find a canoe or something, cross the five-mile strait to Komodo and then all we have to do is to catch our dragon.'

Macassar turned out to be a beleaguered town. Most of Sulawesi was in the hands of rebels who occasionally left their mountain headquarters and came down to the outskirts of the town to ambush the lorries which drove between the town and its airport. Soldiers wearing jungle-green and carrying sten-guns and pistols lolled about the airport. We were scrutinized very carefully by the immigration officials before we could leave in convoy with an armed escort to spend the night in the town. Next day Charles, Sabran and I returned to the airport, boarded a twelve-seater plane and flew off again, heading south-east. Tiny islet after islet passed beneath us, each no more than a hump of land covered in scorched brown grass, spotted with green palm trees and girdled with white coral-sand beaches. Beyond the irregular lines of surf, the coral shallows sparkled a mottled green until the seafloor fell abruptly beyond the reef and the colour of the sea reverted to brilliant peacock-blue. One after the other the islands floated behind us, each looking exactly the same, and each, I felt sure, virtually indistinguishable from Komodo. But none of these patches of land harboured the giant lizards; Komodo and its satellites are the only islands in the world where they exist.

We droned onwards through the cloudless sky, hanging between the topaz-yellow sun and the sea. After two hours, bigger mountains than we had seen before materialized on the hazy horizon ahead. Flores. As we lost height, the coral-carpeted sea slid beneath us with increasing speed. Ahead we could see angular volcanic mountains. We skimmed over the coastline, over a group of thatched huts clustered round a large white church, until with a shudder our wheels touched the grassy runway.

A white-painted building was the only sign that the grassy field on which we had landed was a regulation airstrip. In front of it stood a group of people watching our arrival, and drawn up beside it – our hearts leaped with relief – we saw a lorry with two men sitting on its front bumper. With the other passengers we followed the pilot and co-pilot to the building. Inside, a dozen sarong-clad men watched us undemonstratively. Physically they were quite different from the smaller straight-haired people we had known in Java and Bali; they had frizzy hair and broader noses, and were more akin to the people of New Guinea and the South Seas. One girl, wearing a forage cap and, unaccountably, a heavy tartan skirt, appeared to be the airline representative.

She immediately began filling in forms with the aircrew. No one rushed forward to meet us. Our luggage arrived on a trolley and was unloaded on to the floor. We hovered above it, hoping that That Sen would identify us from its prominent labels.

'Good afternoon,' I said loudly in Indonesian to everyone in general. 'Tuan That Sen?'

The boys leaning against the walls switched their pensive gaze from our equipment to Charles and me. One of them giggled. The girl in the tartan skirt hurried out on to the airstrip, brandishing her papers.

The men continued to look at us abstractedly until one of them declared himself the customs officer by putting on a peaked cap.

He pointed to our baggage. 'Yours, tuan?'

I smiled extravagantly and began the Indonesian speech I had been rehearsing to myself on the plane.

'We are Englishmen, from London. Unhappily we speak little Indonesian. We have come to make a film. We have many papers. From the Ministry of Information in Jakarta; from the Governor of the Lesser Sunda islands in Singaraja; from the Indonesian Embassy in London; from the British Consul in Surabaya.'

As I mentioned each authority, I handed over a letter or a pass. The Customs Officer received them like a hungry man presented with a delectable meal. As he was digesting them, through the open doorway bustled a fat, perspiring Chinese man. He held out both hands to us, smiled broadly and at the top of his voice burst into a flood of very rapid Indonesian.

I understood the first few sentences, but he spoke with such speed that I soon lost track of what he was saying. I tried twice to stem the flood with a few remarks of my own ('We are Englishmen, from London. Unhappily we speak little Indonesian.') without any effect, so while he talked incomprehensibly, I stared at him, fascinated. He was wearing creased, baggy khaki trousers and shirt and, as he spoke, he continually mopped his brow with a red spotted handkerchief. It was his forehead which interested

me most, for he had shaved an ample three inches of the front
of his scalp. This had the effect of giving him a much deeper
brow than Nature had intended and I became absorbed in trying
to reconstruct his original appearance. His black toothbrush-stiff
hair would obviously have approached to within an inch of his
luxuriant eyebrows. I was brought back from this speculation
with a jerk, for he had finished speaking.

'We are Englishmen,' I said hastily, 'from London. Unhappily
we speak little Indonesian.'

By this time, the customs man had finished his scrutiny of
our pile of documents and was scrawling chalk marks all over
our baggage. That Sen beamed. 'Losmen,' he cried. Seeing that
I didn't immediately understand, he resorted to the classic British
way of dealing with the uncomprehending foreigner and assumed
that I was deaf.

'Losmen!' he bawled in my ear.

This time I recalled the meaning of the word and together
we took our luggage outside to the lorry, which, it appeared,
did indeed belong to That Sen. As we rattled into the town, we
were forced to sit in silence for the noise of the lorry made
speech impossible.

The *losmen* was similar to the guest-houses we had known
elsewhere in Indonesia – a series of dark cement boxes fronted
by a long veranda, each box containing a rectangle of boards,
with a thin rolled mattress at one end, which was to serve as a
bed. We unloaded our luggage and returned to That Sen.

It took us an hour of work with the dictionary to understand
the situation. The only transport in Maumere in working order
was That Sen's lorry, which itself had only just been made road-
worthy so that it could undertake a journey to Larantocha, a
village twenty miles away to the east, the opposite direction to
that in which we wanted to go. There it would be cosseted
for a week to prepare it for the return journey to Maumere.
This was a vital cog in the island's transport system and to com-
mandeer it was unthinkable. That Sen smiled, slapped me on

the back and said, 'No worry about lorry.' Charles, Sabran and I looked at one another gloomily. 'No *worry*,' repeated That Sen. 'I get better idea. Coloured lakes of Flores. Very famous. Very beautiful. Very close. Forget lizards. Film lakes.'

We brushed this suggestion aside. The only other route to Komodo was by sea; was there perhaps a small motor vessel in Maumere harbour? That Sen shook his head vigorously. Perhaps then, a little fishing prau? 'Maybe,' said That Sen, and before we could thank him properly for all his kindness and patience, he had driven off in his quivering lorry to find a boat for us.

It was late in the evening before he reappeared. He bustled out of his lorry and, mopping his forehead, smilingly explained that all was well. The fishing fleet was at sea but, by a stroke of good fortune, one prau was still in harbour and he had brought its Captain with him to discuss our plans. The Captain was a rather shifty-looking man wearing a sarong and a black *pitji*. He allowed That Sen to do all the talking, nodding his approval or disagreement, his eyes fixed mostly on the floor.

The trade winds, we knew, were blowing from Maumere towards Komodo and we suggested that if the Captain would take us there, we should then continue westwards with the wind behind us to Sumbawa, where we could hope to catch another plane. The Captain nodded. All that remained to be settled was the price. We were hardly in a position to bargain, for both That Sen and the Captain knew that we were determined to go to Komodo and that without the Captain's boat we should never get there. The price finally agreed was an extremely high one. The Captain left well pleased, having said that he would be ready the next day.

We had many things to do. Maumere police station had to be visited, the customs at the harbour had to be placated, and our return air-tickets had to be cancelled. Lastly, we called on That Sen in his store to buy our provisions for the journey. We had to limit our purchases, as the Captain's fee would absorb most of our money and we had to keep some in reserve, for we had no notion of what financial emergencies we might have to face before we got back to Java. We bought a few luxuries – some tins of corned beef and condensed milk, some dried fruit, a large can of margarine, several bars of chocolate – but our main purchase was a large bag of rice, for That Sen assured us that on rice, augmented by the great quantities of fresh fish the Captain would catch for us on the voyage, we should be able to live for weeks and weeks.

We arrived in the harbour in the late afternoon with all our purchases. The Captain was not there, but That Sen introduced us to the ship's crew, two boys about fourteen years old, Hassan and Hamid. They, like the Captain, were straight-haired and sharp-featured. They wore checked sarongs which, as they helped us take our luggage on board, they hitched up to reveal scarlet bloomers.

The prau was even tinier than we had expected. Twenty-five feet long, it was single-masted and carried a triangular mainsail swinging on a bamboo boom and a small foresail which was also attached to a boom by its lower edge. Behind the mast, and abutting on it, stood a low ridge-roofed cabin. The top of its roof was no more than three feet above the deck and we could only enter by crawling on our hands and knees. It was floored by a mat of split bamboo, lying over three cross-timbers. This mat rolled back and exposed the gaping hold. Piece by piece, we handed our gear through the cabin and into the hold, laying it on the piles of coral rock in the ship's bottom which were there to act as ballast and which served as a platform to keep our equipment above the dirty water slopping about in the bilge. The stench below was overwhelmingly foul, a mixture of stale brine, cola nuts and putrescent salted fish, and we were glad when everything was loaded.

It was late in the afternoon before the Captain reappeared. That Sen stood on the jetty, still mopping his brow. We thanked him again and again, Hassan and Hamid hoisted the sails, and with the Captain at the tiller we cast off.

It was a fine evening, the wind was fresh and strong and the little boat thrust forward eagerly through the choppy sea. Charles elected to sleep on the foredeck; Sabran and I bedded down on the bamboo mat in the cabin together with Hassan and Hamid. It was difficult to decide which was the pleasanter berth. Charles ran the risk of being wakened by a shower of rain and of being clouted over the head by the foresail boom which swung over him, a foot above his face, every time the ship tacked. On the other hand, he had the fresh air, which was more than could be said for anyone who jammed his body inside the cabin which was filled with the smell of rotten fish from the hold below. However, none of us was inclined to grumble; we were, after all, on our way.

Sabran preparing rice on the prau

When I woke, I realized from the movement of the boat that the wind had dropped. Through the cabin doorway I could see the Southern Cross sparkling in a cloudless sky. Then I heard again the noise which had woken me, a horrifying crunch which made the ship shiver and lurch. I scrambled through the doorway on to the deck. Charles was already awake and peering over the side.

'We,' he announced dispassionately, 'are on a coral reef.'

I yelled at the Captain, huddled over the tiller. He didn't move. I clambered quickly aft and shook him. He opened his eyes reproachfully. '*Aduh*, tuan,' he said. 'Not do that.'

'Look,' I cried excitedly, pointing over the side as the ship shook with another crunch.

The Captain tapped his right ear. 'This one no good,' he said aggrievedly. 'I not hear well.'

'We are on a reef,' I shouted in desperation. 'That no good either.'

The Captain wearily got to his feet and woke Hassan and Hamid. Together they pulled out a long bamboo pole lying along the ship's side and began pushing us off the reef. The moon was bright enough for us to distinguish the plates and knobs of coral a few feet below the surface. The water was laced with bright lines of phosphorescence, and every time the gentle swell lifted the boat and ground it on to the coral the water became suffused with a greenish glow.

Ten minutes' work and we were rocking gently in deeper water. The boys returned to the cabin and lay down; the Captain curled himself up in his sarong beside the tiller and went back to sleep.

The incident disturbed Charles and me. Perhaps there had been no danger as we were rolling in a dead calm, but whenever I had read of travellers going aground on a coral reef the incident had always ended in disaster. I felt a little unnerved and my confidence in the Captain was somewhat shaken. It was difficult to return to sleep, so we sat on deck talking for an hour or more. Dimly on the horizon we could distinguish the shadowy form of a large island. The sails flapped idly above us. The boat rose and fell with the swell. A little chi-chak gecko somewhere up the mast suddenly called. At last we fell asleep again.

When we awoke, the island we had seen during the night was in exactly the same position as it had been six hours before. We had not moved an inch. All that day we lay becalmed, slowly pivoting on the blue glassy water. We sat and smoked, casting our cigarette ends over the side where they remained, so that in the evening the still water in which we lay was covered with an increasing accumulation of our own litter. We glared at the island ahead of us. Hassan and Hamid slept. The Captain lay back by the tiller, his hands behind his head, staring vacantly at the sky. Occasionally he absentmindedly let out a loud falsetto yell. It was, I suppose, a sort of song, but after a few hours we found it a little irritating. The day dragged by and we settled down for another night. In the morning the island was still in exactly the same position. We hated the sight of it. All day we lay on board waiting for a puff of wind to stir the sails which hung limply from the rigging. Yesterday's cigarette ends still floated dismally a few feet away. Charles and I sat in the broiling sun dangling our feet over the side in the tepid water. Sabran occupied himself with cooking. Our only fresh water on board was contained in a large stoneware jar, lashed to the wooden wall of the cabin. Although its mouth was covered with a small pottery dish, it was nevertheless full of wriggling mosquito larvae. The sun, as we lay becalmed, smote us so hard that the jar was already uncomfortably hot to touch and the water inside it unpleasantly warm. Sabran made it both palatable and innocuous by boiling it, dissolving several sterilizing tablets in it, and adding sugar and coffee powder. Fortunately the heat made us so thirsty that we were not over-particular as to what we drank. My appetite quailed, however, when he produced the fourth successive meal of plain, ungarnished, boiled rice.

I clambered aft to talk with the Captain who lay in the stern singing spasmodic snatches of his chant to himself.

'Friend,' I said. 'We very hungry. You catch fish now?'

Charles filming on board the prau

'No,' replied the Captain.

'Why not?'

'No hooks. No line.'

I was indignant. 'But Tuan That Sen say you fisherman!'

The Captain hitched up the right side of his mouth in a wet sucking sniff.

'Not,' he said.

This was not only a severe blow to our catering arrangements but something of a mysterious statement. If he was not a fisherman, what was he? I pestered him with further questions, but could get no more information from him.

I went for'ard and joined Charles in eating a saucepanful of plain, ungarnished, boiled rice.

After the meal, Charles and I took refuge from the scorching sun in the cabin. We lay semi-naked on the hard bamboo slats, sweating uncomfortably. I was roused from my torpor by a distant puffing and snorting. Poking my head outside I saw three hundred yards away a large school of dolphins. An area of the sea the size of a football pitch was whitened and flecked with their splashes as, filled with the exuberance which seems to possess all dolphins, they leaped from the waves and curvetted through the air. Some, less energetic, broke the surface of the water with their foreheads only and cleared their lungs through their blowholes with the loud snorts which had aroused me. When we first saw them they were travelling obliquely towards us as we lay motionless in the calm, but as we watched they visibly changed their course to come and inspect us. Within a few seconds the school was all around us. We hung over the side, peering at them through the shifting green translucencies of the water as they cavorted around our bows. They came so close, swooping through the water, that we could see every detail of their bodies; their beak-like mouths, the large dark blowholes in their foreheads, and their humorous eyes with which they, on their part, looked quizzically back at us.

For perhaps two minutes they gambolled about us. Then, puffing and splashing, they moved off towards the island on the horizon ahead. We followed them with regretful eyes and once again we were enveloped by the stillness of the sea. As evening drew in, our sails flapped slightly. Overboard, I noticed that the trail of paper and cigarette ends was lying a considerable distance astern. Soon the breeze strengthened into wind, and as the sun neared the horizon we started to plunge and bucket through the roughening sea. Big waves began to chase and overtake us. As each one caught us, it lifted the little ship's stern so high that she dipped her bowsprit deep into the water ahead; as it passed her, so she reared back, lifting her dripping foresail to the sky. That night, when I lay down to sleep in the cabin, the forked pivot of the heavy boom was swinging and straining against the mainmast, braying like a drunken trombonist. It was the sweetest noise I had heard for a long time.

The wind stayed strong and fair throughout the next day. To

port, the coast of Flores lay like a long ribbon along the horizon. Shoals of flying fish crossed our bows. They rose in the body of a wave until, before it broke, they burst from its crest, unfurled their blue and yellow pectoral fins, and took to the air. They glided as far as twenty yards in a single flight, dodging between the waves with swift jerky turns. When a large school passed us, the moving troughs of the waves were alive with the skittering twisting flight of these beautiful creatures.

Our consciences reminded us that we should be recording the voyage on a film. Charles clambered down into the hold and assembled his camera. The Captain provided an obvious subject as he squatted on the deck, leaning sleepily on the tiller, his sarong draped over his head to protect him from the sun. The foam-flecked sea rose and fell behind him as the prau plunged through the waves.

'Friend,' I said. 'Photo?'

He came to life with a start.

'No, no!' he said aggressively. 'No photo. Not agree!'

The mystery of the Captain's character deepened. He was the first Indonesian we had met who was not more than anxious to have his photograph taken. We had obviously been tactless, though why I could not imagine. I tried to repair the breach of etiquette, if such there was, by gossiping while Charles looked for other subjects for his camera.

'Good wind,' I said, conversationally, looking at the sails swelling white against the cloudless blue sky.

The Captain grunted, narrowed his eyes and glared straight ahead.

'We reach Komodo tomorrow with such wind?' I asked.

'Maybe,' the Captain replied.

He paused, sniffed and then emitted one of his falsetto yells. I took this to imply that our conversation was ended and returned for'ard.

That night was our fourth at sea. I reckoned that we must be nearing Komodo, and when I woke in the morning I fully expected to be greeted with the news that we were within sight. I scanned the horizon anxiously. All that could be seen was the continuous bumpy line of Flores stretching along the southern horizon.

I found the Captain dozing inside the dugout canoe which lay on its side on the deck alongside the cabin.

'Friend,' I said. 'How many hours to Komodo?'

'Not know,' he replied sulkily.

'Have you been to Komodo before?' I said, trying to sting him into making some sort of estimate.

'*Belum*,' said the Captain.

This was a new word to me. I crawled into the cabin to find my dictionary.

'*Belum:* not yet,' I read. An awful suspicion formed in my mind. I crawled out of the cabin: the Captain had gone back to sleep.

I shook him gently.

'Do you know where Komodo is, Captain?' I asked.

He shifted himself into a more comfortable position.

'I not know. Tuan know.'

'Tuan', I said loudly and with decision, 'does *not* know.'

He swung himself up into a sitting position.

'*Aduh!*'

I went back to the cabin, rummaged in my kit for maps and called Charles away from filming picturesque close-ups of the sails. We had two maps. One was a large and necessarily simplified one of the whole of Indonesia which I had begged from the shipping company. This showed Komodo as a tiny blob less than an eighth of an inch long. The other was a very large detailed map of Komodo itself, which I had sketched from a scientific monograph which showed the island and its surrounding islets in enormous detail, but only the very tip of Flores. It would be of little use to us until we were within sight of Komodo.

We showed the Captain the map of Indonesia.

'Where does the Captain think we are?'

18

The Island of Komodo

As the first glimmerings of dawn spread across the sea, I uncurled my stiff limbs and picked myself up from the deck on which I had slept during the last three hours of the night. Charles and Hassan, still on watch, were leaning drowsily against the cabin roof with bamboo poles at the ready, although the tide was already slackening and there was no longer any real danger of our boat being swept on to the rocks that had menaced us the night before. Sabran made his way for'ard, carrying a pot of boiling, salty, chlorinated coffee. As we sipped it gratefully, the rim of the sun bulged above the horizon behind us, warming our semi-naked bodies. Ahead of us, its rays illuminated three jagged islets which stood like a screen in front of a line of more distant hazier mountains. To the left, two miles away, a coastline, distinguished by an almost symmetrical pyramidal mountain, stretched towards the three islets but subsided into the sea just before it met them, leaving a narrow gap which, we presumed, must be the gateway to the Indian Ocean. The land to the right, in the lee of which we had sheltered, was, we hoped, Komodo. This was our first sight of it in daylight, and I scanned the steep grassy slopes which rose above us, half hoping to see the scaly head of a dragon peer from behind one of the rocks.

There was no wind. We all took bamboos and slowly poled the ship out of the bay. In the centre of the strait the tide was still running strongly, and with no wind to give us motive power we dared not venture into deeper water, so we continued to inch our way round the coast by poling. From our map, Charles and I were sure that the screen of islets guarded and concealed the entrance to Komodo Bay. When we were still a mile away from the islets, the water shallowed and our keel grated on the bottom. The boat could go no further until the tide rose again and lifted us.

To sit still for three hours and wait for this to happen, however, was not to be borne. Leaving Charles with the Captain and crew, Sabran and I clambered into our small dugout canoe and paddled on ahead to see if we could confirm the existence of the bay behind the islets.

We kept close inshore. The coral lay thick beneath us and often our canoe cleared it by only a few inches. Occasionally, great rocky hummocks of brain coral projected to within an inch of the surface. If we had hit one, the little canoe would certainly have capsized and we should have been thrown half-naked into the stony forest of jagged staghorn coral. Sabran, however, was a master canoeist. He spotted the dangers well ahead and with a flick of his paddle deflected the dugout on to a safe course. As we paddled, thin elongated fish about twelve inches long leaped from the water ahead of us in groups of two or three and skittered along the surface. Their bodies were inclined at forty-five degrees to the surface of the sea and only the tips of their tails remained in the water vibrating rapidly, driving them along, until after they had travelled several yards they fell forward on their bellies and disappeared.

We reached the trio of islets. As we passed between the right-hand one and the mainland, a large beautiful bay opened in front of us. It was girt by bare mountains, steep, gaunt and fawn brown. At the far side, a slim curve of white, rimming the amethyst water of the bay, indicated a sandy beach. Above it, at the base of the hills, we saw a patch of dark green. This we guessed to be a grove of palm trees, which might shelter a village. We paddled eagerly across the deeper waters of the bay. Soon we were able to distinguish outrigger canoes lying on the beach and a few grey thatched huts among the palms. At last we had definite proof that our navigation during the past few days had been correct. This must be Komodo, for Komodo is the only island in the whole group which is inhabited.

A few naked children were standing on the beach watching us as we hauled our canoe up on to the sand. We walked across the coral and shell-strewn beach, towards the wooden atap-thatched huts which stood on stilts in a line between the beach and the steep hillside behind. In front of one of the huts an old woman squatted on her haunches taking shrivelled fragments of shellfish from a basket and laying them carefully in long rows on a strip of coarse brown cloth spread on the sand beside her, so that they would dry in the baking sun.

'Peace on the morning,' I said. 'The house of the *petinggi*?'

She brushed her long greying hair from her wrinkled face, screwed up her eyes, and without showing any sign of surprise at seeing two total strangers in her village pointed to a hut farther along the line which was a little larger and less decrepit than its neighbours. We walked across to it, the sand hot on our bare feet, watched only by children and a few old women. The *petinggi* stood in the doorway of his hut awaiting us. He was an old man wearing a smart clean sarong, a white shirt, and a black *pitji* placed squarely on his brow. He gave us a broad but completely toothless smile, shook both of us by the hand and invited us into his house.

As we entered we realized why the village had seemed semi-deserted; the room was crowded with men squatting on the rattan mats which covered the floor. The room itself, perhaps five yards square, was devoid of any furnishing except for a large ornate wardrobe with a cracked and mottled mirror attached to its door. Three of the walls of the room were of wood, the fourth, facing the entrance, was a plaited palm-leaf screen. A length of grubby cloth hung down by the edge of the screen concealing a passage leading to the other half of the hut where, as I later learnt, the cooking was done. Four young women peered round the edges of this curtain, looking at us with saucer eyes. The *petinggi* gestured to us to sit down in a small clear space in the centre of the floor. The curtain was pushed to one side and one of the women shuffled in, picking her way with difficulty through the seated men with her body bent double in a symbolic endeavour to keep her head below that of the menfolk, a traditional sign of respect. She carried a plateful of fried coconut cakes which she set in front of us. Another woman followed with cups of coffee. The *petinggi* sat down, cross-legged, facing us, and together we ate and drank. After we had completed our lengthy greetings, I explained as best I could who we were and what we had come for. When I was lost for a Malay word, I turned for help to Sabran, sitting at my side. Nearly always he had anticipated what I wanted to say and was able to prompt me immediately. Sometimes, however, to find the right word we had to have a hurried consultation in our own private vocabulary of signs and ill-pronounced words. Sometimes, too, I produced a newly acquired word or phrase without prompting, and then Sabran beamed with delight and whispered in English 'Is very OK.'

The *petinggi* nodded and smiled throughout. I handed round cigarettes. After half an hour, it seemed to me that I might begin to talk of practicalities and to mention our boat lying stranded in the shallows several miles away.

'Tuan,' I said, 'perhaps a man is able to come back to our prau to show the way through the reefs?'

The *petinggi* smiled and nodded assent. 'Haling, my son, will do so.'

However, it was clear that he did not consider that there was any urgency, for more cups of coffee were produced.

The *petinggi* changed the subject.

On our way to the active volcano, Bromo, in East Java.

With Charlie, the orangutan, on board the *Kruwing*.

When we first encountered Benjamin the bear cub, he needed to be fed every three hours.

Hassan at the tiller of the prau, en route to Komodo.

Becalmed, we waited for days for a puff of wind to send us on our way.

The largest of the Komodo dragons seemed completely unconcerned by our presence.

A butterfly swarm at Ihrevu-qua.

Charles in a patch of *monte* in the scrubland of the Paraguayan Chaco.

The giant armadillo – the animal that had eluded us completely –
meeting one of its smaller cousins at the London Zoo.

'I sick,' he said, extending his left hand which was badly swollen and smeared with white mud. 'I put on mud as medicine but it not get better.'

'Plenty, plenty good medicine on prau,' I said, hoping to steer his thoughts back to our immediate problem. He nodded gently and asked to look at my watch. I unstrapped it and handed it to him. He examined it and passed it round to the other men who looked at it admiringly and held it to their ears.

'Very good,' said the *petinggi*. 'I like.'

'Tuan,' I replied, 'I am not able to give. Watch is a present to me from my father. But,' I added pointedly, 'we have gifts on the prau.'

Another round of coconut cakes appeared.

'You take photo,' asked the *petinggi*. 'One time, Frenchman here. He take photo. I like very much.'

'Yes,' I replied. 'We have a camera on our prau. When it comes here, we will take photo.'

At last the *petinggi* considered that the conference had proceeded long enough for the requirements of custom to be satisfied. Everyone trooped out of the hut on to the beach. He pointed to an outrigger canoe lying on the sand.

'The boat of my son,' he said and with that he left us.

Haling was meanwhile changing from his best sarong into more workmanlike gear, and at last, two hours after Sabran and I had landed, we helped him push his canoe into the sea. Half a dozen other men joined us. We erected a bamboo mast, hoisted an oblong loosely woven sail and with a stiff following breeze we scudded over the choppy sea towards our boat.

When we arrived, we found that Charles had not been idle while we had been away but had been filming views of the island. His cameras and lenses lay strewn on the foredeck. The Komodo men scrambled on board and seized them with delight. Hastily we explained that unhappily we could not permit this and as quickly as possible we packed away our equipment and

Returning to our prau

stowed everything in the hold. The men, baulked, migrated to the stern. When we rejoined them, they were sitting talking to the Captain and Hassan. Haling was holding our precious tin of margarine, which, when I had last seen it, had been three-quarters full.

He scraped the bottom with his fingers, brought out a lump of margarine and smeared it on his long black hair. I looked round at the rest of the men. Some were massaging their scalps, others merely licking their fingers. The tin, I could see, was empty; we had lost our entire supply of cooking fat; we could not even vary our diet of boiled rice with fried rice.

I nearly said something in anger, but there was little point; it was too late.

Instead, Haling spoke.

'Tuan,' he said, rubbing his scalp with his greasy fingers, 'you have comb?'

That evening, the boat anchored safely in the bay, we sat in the *petinggi's* hut discussing our plans in detail. The *petinggi* called the giant lizards *buaja darat* – land crocodiles. There were very many on the island, he told us, so many that sometimes one would wander right into the village to scavenge among the refuse tips. I asked him whether anyone from the village ever hunted them. He shook his head vigorously. *Buaja* were not so good to eat as the wild pigs which were abundant, so why should his men kill them? And in any case, he added, they were dangerous animals. Only a few months ago, a man was walking through the bush when he stumbled upon a *buaja* lying motionless in the *alang-alang* grass. The monster had struck with its powerful tail, knocking the man over and numbing his legs so that he was unable to escape. The creature then turned and mauled him with its jaws. His wounds were so severe that he died soon after his comrades found him.

We asked the *petinggi* how we could best attract the lizards so that we could take photographs. He was in no doubt. They have a very keen sense of smell, he told us, and they will come from very long distances to putrefying meat. He would slaughter two goats that night and tomorrow his son would take them to a place on the other side of the bay where the *buaja* were plentiful. All would be well.

The night was a clear one. The Southern Cross sparkled in the sky above the toothed silhouette of Komodo. Our prau rocked gently in the calm waters of the bay. We had just finished a meal which for the first time in six days did not contain a grain of rice, for Sabran had managed to obtain two dozen tiny chicken's eggs in the village and had made us a gigantic omelette which we had washed down with cool slightly fizzy coconut milk. Charles and I lay on the foredeck with our heads pillowed on our hands watching the unfamiliar constellations wheel across the sky. Several times giant shooting stars blazed an incandescent trail across the black bowl above us. The steady beat of gongs drifted over the inky waters from the village. My mind kept turning again and again to what might await us in the morning. I was so excited that I didn't fall asleep until late in the night.

———

We rose at dawn and ferried all our equipment ashore in the dugout. I had hoped to be off early, but it took nearly two hours for Haling to prepare himself for the voyage and to collect three other men to help carry all our gear. At last, we helped him push his fifteen-foot-long outrigger canoe down the beach and into the water. We loaded it with our cameras, tripods and recording machines, together with the two goat carcasses slung on a long bamboo.

The sun had already risen above the brown mountains ahead of us as we set out across the bay. The water spurted white over the bamboo outrigger. Haling sat in the stern holding the rope attached to one corner of the rectangular sail, adjusting its trim to suit the varying wind. Soon we were sailing beneath steep rocky cliffs. High up on a ledge above us a splendid fish eagle stood alert, the sun glinting on the chestnut feathers of his mantle.

We landed at the mouth of a valley choked with scrub which ran down from the bare grass-covered mountains. Haling led us inland, cutting a path through the thorn bushes. We walked for an hour. Occasionally we crossed patches of savannah, open grassy areas containing a few top-heavy lontar palms, their thin columnar trunks rising branchless for fifty feet before bursting into an explosion of feathery leaves. Here and there we passed a dead tree, its barkless bleached branches riven by the heat of the sun. There was no sign of life except for the chirring of stridulating insects and the vociferous screams of parties of sulphur-crested cockatoos which fled ahead of us. We waded across a muddy brackish lagoon and continued our way through the bush towards the head of the valley. It was oppressively hot; a blanket of lowering clouds had spread across the sky, seemingly preventing the heat from leaving the baked land.

At last we came to the dry gravelly bed of a stream, as wide and as level as a road; on one side it was overhung by a bank fifteen feet high which was draped with a tangle of roots and lianas. Tall trees grew above, arching over the bed to meet the branches of the trees from the other bank and forming a high spacious tunnel down which the streambed curved and disappeared.

Haling stopped and put down the equipment he had been carrying. 'Here,' he said.

Our first task was to create the smell which we hoped would attract the lizards. The goats' carcasses, already decomposing slightly in the heat, were blown up and swollen as tight as drums. Sabran slit the underside of each one and a foul-smelling gas hissed out. Then he took some of the skin and burnt it on a small woodfire. Haling climbed one of the palms and chopped down a few leaves with which Charles improvised a hide, while Sabran and I staked out the carcasses on the gravelly streambed fifteen yards away. This done, we retired behind the palm-leaf screens and began our wait.

Soon it began to rain, the drops pattering gently on the leaves above us. Haling shook his head.

'No good,' he said. '*Buaja* no like rain. He stay in his room.'

As our shirts got wetter and the rain water trickled down the channel of my back, I began to feel that the *buaja*, on the whole, was more sensible than we were. Charles sealed all his equipment in watertight bags. The smell of the rotting goats' flesh permeated the air. Soon the rain stopped and we left our hides beneath the dripping trees and sat on the open sandy bed of the stream to dry. Haling gloomily insisted that no *buaja* would leave their lairs until the sun shone and a breeze sprang up to disperse the stench of the goats' meat which hung around us. I lay back miserably on the soft gravel of the streambed and shut my eyes.

When I next opened them, I realized with surprise that I must have fallen asleep. I looked round and saw that not only

Awaiting the arrival of the dragon

Charles, but Sabran, Haling and the other men were also fast asleep, their heads resting in one another's laps or on our equipment boxes. I felt that perhaps it would have served us right if the lizards had come out in spite of the rain, and consumed all our bait. But the goats lay there untouched. I looked at my watch. It was three o'clock. Although it had stopped raining there was still no sign of a break in the clouds and it seemed very unlikely that the lizards would come to our bait that day. Time, however, was precious. At least we could build a trap which we could leave overnight, so I roused everyone.

During the past few weeks, Charles and I had often discussed with Sabran the best type of trap to use for catching our dragon. Eventually we had decided upon one which Sabran employed to catch leopards in Borneo. Its great merit was that, apart from a length of stout cord, all the materials needed to make it could be obtained from the forest.

The main body of a trap, a roofed rectangular enclosure about ten feet long, was easily constructed. Haling and the other men began to cut strong poles for us from the trees growing on the banks, and Charles and I selected four of the stoutest and drove them into the streambed, using a big boulder as our pile-driver. These were our corner posts. Sabran meanwhile had climbed a tall lontar palm and cut down several large fan-shaped leaves. He then took the stems, split them, crushed them and thrashed them on a boulder to make them pliable. After he had given the resulting fibres a twist, he handed them to us – pieces of strong serviceable string. With them, we lashed long horizontal poles between our corner posts, strengthening the structure with uprights where we thought it necessary. At the end of half an hour we had built a long enclosed box open at one end.

Now we had to make a drop-door. This we built of heavy stakes tied together with Sabran's string. The vertical ones were sharpened at the bottom so that when the door fell they would stick deep into the ground, and the lowest horizontal cross-piece overlapped the corner posts inside the trap so that when the door fell it could not be pushed outwards by the dragon – should we ever get one inside. We completed the door by tying a heavy boulder on it with lianas so that it would be difficult to lift once it had fallen.

All that remained to do was to build the triggering device. First we pushed a tall pole through the roof of the trap and drove it into the ground near the enclosed end. Then we planted the feet of two more on either side of the door and tied them in a cross, directly above it. We knotted our cord on to the door, raised it, and ran the cord over the angle of the cross-poles to the upright post at the other end of the trap. Instead of keeping the door raised by tying the cord directly on to the post, we tied it on to a small piece of stick about six inches long. Holding the stick upright and close to the post, we twisted two rings of creeper round it and the post, one near its top and one near its bottom.

The dragon trap

The weight of the door pulled the cord tight and so prevented the rings from slipping down the pole. Then we fastened a smaller piece of cord to the bottom ring, threaded it through the roof of the trap and attached it to a piece of goat's flesh inside.

To test it, I poked a stick through the bars of the cage and jabbed the bait. This tugged the cord attached to the bottom ring and pulled it downwards. The small piece of stick flew loose and the door at the other end dropped with a thud. Our trap worked.

Two last things remained to complete it. First we piled boulders along the sides so that a dragon inside the trap would not be able to insert its nose under the lowest poles and uproot the entire construction. Then we shrouded the closed end with palm leaves so that the bait could only be seen through the open door.

The three of us then dragged the remainder of the goats' carcasses to the foot of a tree, threw a rope over a projecting branch and hauled them into the air so that they would not be eaten during the night and their smell would spread widely through the valley, attracting dragons to the trap.

We gathered up all our equipment and walked back in the drizzling rain to the canoe.

That night the *petinggi* entertained us in his hut. We squatted on the floor drinking coffee and smoking cigarettes. The headman was in a thoughtful mood.

'Women,' he said. 'How much in England?'

I was unsure how to reply to this.

'My wife,' he added mournfully, 'cost me two hundred rupiahs.'

'*Aduh!* In England sometimes when a man marries a woman, her father will give *him* much money!'

The *petinggi* was astonished. Then he assumed an air of mock seriousness.

'Not tell this to men in Komodo,' he said gravely. 'They all get in canoes and sail straight to England.'

The conversation turned to the prau in which we had sailed to Komodo and in particular to the Captain. We told the *petinggi* of our difficulties in reaching his island.

He snorted.

'That Captain. He no good. He not a man of these islands.'

'From where does he come?' I asked.

'From Sulawesi. He carried guns from Singapore and sells to rebel army in Macassar. Government men, they find out, so Captain, he sails down to Flores and not go back.'

This explained a great deal – the lack of fishing tackle, the ignorance of the whereabouts of Komodo, and the Captain's reluctance to have his photograph taken.

'He say to me,' added the *petinggi* as an afterthought, 'maybe men from this *kampong* sail with you when you leave.'

'Of course. We are very happy to have them. Do they wish to go to Sumbawa?'

'No,' said the *petinggi* airily. 'But Captain say you have much money, and many valuable things. He say, maybe if there are men to help him, he can get them from you.'

I laughed lightly but a little nervously. 'Are they coming?'

He looked at me pensively.

'I not think so,' he replied. 'We have much fishing to do here you know, and they not wish to leave their families.'

19

The Dragons

———

The sky was cloudless as we sailed across the bay early the next morning. Haling, seated in the stern of the outrigger canoe, smiled and pointed to the sun, which was already shining fiercely.

'Good,' he said. 'Much sun; much smell from goats, many *buaja*.'

We landed and set off through the bush as fast as we could. I was impatient to go back to the trap, for it was just possible that a dragon might have entered it during the night. We pushed our way through the undergrowth and emerged on to one of the patches of open savannah. Haling was ahead when suddenly he stopped. '*Buaja*,' he called excitedly. I ran up to him and was just in time to see, fifty yards away on the opposite edge of the savannah, a moving black shape disappear with a rustle into the thorn bushes. We dashed over to the spot. The reptile itself had vanished but it had left signs behind it. The previous day's rain had collected in wide shallow puddles on the savannah, but the morning sun had dried them, leaving smooth sheets of mud, and the dragon we had glimpsed had walked over one of these leaving a perfect set of tracks.

Its feet had sunk into the mud, its claws leaving deep gashes. A shallow furrow, swaying between the pug-marks, showed where the beast had dragged its tail. From the wide spacing of the footprints and the depth to which they had sunk in the mud, we knew that the dragon we had seen had been a large and heavy one. Although our view of the monster had been so short

251

and fragmentary, we were very excited by it; at last we had seen for ourselves the unique and wonderful creature which had been dominating our thoughts for so many months.

We delayed no longer over the tracks, but hurried on towards the trap through the dense bush. As we reached a tall dead tree, which I recognized as being within a very short distance of the streambed, I was tempted to break into a run, but I checked myself with the thought that to crash noisily through the bush so close to the trap would be a very foolish thing to do, for a dragon might at that very moment be circling the bait. I signalled to Haling and the men to wait. Charles, gripping his camera, Sabran and I picked our way silently through the undergrowth, placing each foot with care and circumspection lest we should tread on a twig and snap it. The shrill cry of a cockatoo rang out above the chirring of insects. In the distance it was answered by a briefer more savage crowing birdcall.

'*Ajam utan*,' whispered Sabran. 'Jungle cock.'

I parted the dangling branches of a bush and peered through across the clear emptiness of the riverbed. The trap stood a little below us, a few yards away. Its gate was still hitched high. I felt a wave of disappointment and looked round. There was no sign of a dragon. Cautiously we clambered down to the riverbed and examined the trap. Perhaps our trigger had failed to work and the bait had been taken leaving the trap unsprung. Inside, however, the haunch of goat's meat was still hanging, black with flies. The smooth sand around the trap was unmarked except for our own footsteps.

Sabran returned to fetch the boys with the rest of our recording and photographic equipment. Charles began to repair the hides we had erected the day before, and I walked farther up the streambed to the tree in which we had suspended the major part of our bait. To my delight, I saw that the sand beneath was scuffed and disturbed. Without doubt something had been here earlier trying to snatch the bait. I could understand why, for the stench produced by this mass of rotting meat was incomparably stronger than that given off by the small haunch in the trap. The carcasses were covered by a blanket of handsome orange-yellow butterflies, flexing their wings as they fed on the meat. I reflected sadly that natural history often deals harshly with our romantic illusions about wild life. The most brilliantly beautiful butterflies of the tropical rain forest do not fly in search of appropriately gorgeous blooms, but instead seek a meal from carrion or dung.

As I untied the rope and lowered the carcasses, the butterfly carpet disintegrated into a flapping cloud, mingling with a swarm of black flies which buzzed around my head. The smell was almost more than I could stand. These big carcasses were obviously more potent magnets than the bait in the trap, and as our primary task was to film the giant lizards, I dragged the meat to a place on the streambed which was in clear view of the cameras in the hide. Then I drove a stout stake deep into the ground and tied them securely to it, so that the dragons would be unable to pull them away into the bush and, if they wished to eat, would be compelled to do so within easy range of our lenses. That done, I joined Charles and Sabran behind our screen and settled down to wait.

The sun was shining strongly and shafts of light struck through the gaps in the branches above us, dappling the sand of the riverbed. Although we ourselves were shaded by the bush, it was so hot that sweat poured down us. Charles tied a large handkerchief around his forehead to prevent his perspiration trickling on to the viewfinder of his camera. Haling and the other men sat behind us, chatting among themselves. One of them struck a match and lit a cigarette. Another shifted his seat, and sat on a twig which snapped with a noise that seemed to me to be as loud as a pistol shot. I turned round in irritation with my finger to my lips. They looked surprised but relapsed into silence. Once again I looked anxiously through my peephole in the hide, but almost immediately one of the men began talking again. I turned back to them and spoke in an urgent whisper.

'Noise no good. Go back to boat. We come to you there when our work is finished.'

They looked a little injured – perhaps because they knew (and I at the time did not) that the hearing of a Komodo dragon is very limited. Nonetheless, the noise that they made was very distracting, and I was relieved when they got to their feet and disappeared through the bush.

There was now little noise. A jungle cock crowed in the distance. Several times a fruit dove, purple-red above and green below, shot with closed wings like a bullet along the clear channel above the streambed, soundless except for the sudden whistle of its passage through the air. We waited, hardly daring to move, the camera fully wound, spare magazines of film beside us and a battery of lenses ready in the open camera case.

After a quarter of an hour, my position on the ground became extremely uncomfortable. Noiselessly, I shifted my weight on to my hands, and uncrossed my legs. Next to me, Charles crouched by his camera, the long black lens of which projected between the palm leaves of the screen. Sabran squatted on the other side of him. Even from where we sat, we could smell only too strongly the stench of the bait fifteen yards in front of us.

We had been sitting in absolute silence for over half an hour when there was a rustling noise immediately behind us. I was irritated; the men must have returned already. Very slowly so as not to make any noise, I twisted round to tell the boys not to be impatient and to return to the boat. Charles and Sabran remained with their eyes riveted on the bait. I was three-quarters of the way round before I discovered that the noise had not been made by men.

There, facing me, less than four yards away, crouched the dragon.

He was enormous. From the tip of his narrow head to the end of his long keeled tail I guessed he measured about ten feet. He was so close to us that I could distinguish every beady scale in his hoary black skin, which, seemingly too large for him, hung in long horizontal folds on his flanks and was puckered and wrinkled round his powerful neck. He was standing high on his four bowed legs, his heavy body lifted clear of the ground, his head erect and menacing. The line of his savage mouth curved upwards in a fixed sardonic grin and from between his half-closed jaws an enormous yellow-pink forked tongue slid in and out. There was nothing between us and him but a few very small seedling trees sprouting from the leaf-covered ground. I nudged Charles, who turned, saw the dragon and nudged Sabran. The three of us sat staring at the monster. He stared back.

It flashed across my mind that at least he was in no position to use his main offensive weapon, his tail. Further, if he came towards us both Sabran and I were close to trees and I was sure that I would be able to shin up mine very fast if I had to. Charles, sitting in the middle, was not so well placed.

Except for his long tongue, which he unceasingly flicked in and out, the dragon stood immobile, as though cast in gunmetal.

The biggest of the dragons

For almost a minute none of us moved or spoke. Then Charles laughed softly.

'You know,' he whispered, keeping his eyes fixed warily on the monster, 'he has probably been standing there for the last ten minutes watching us just as intently and quietly as we have been watching the bait.'

The dragon emitted a heavy sigh and slowly relaxed his legs, splaying them so that his great body sank on to the ground.

'He seems very obliging,' I whispered back to Charles. 'Why not take his portrait here and now?'

'Can't. The telephoto lens is on the camera and at this distance it would fill the picture with his right nostril.'

'Well, let's risk disturbing him and change lens.'

Very, very slowly Charles reached in the camera case beside him, took out the stubby wide-angle lens and screwed it into place. He swung the camera round, focused carefully on to the dragon's head and pressed the starting button. The soft whirring of the camera seemed to make an almost deafening noise. The dragon was not in the least concerned but watched us imperiously with his unblinking black eye. It was as though he realized that he was the most powerful beast on Komodo, and that, as king of his island, he feared no other creature. A yellow butterfly fluttered over our heads and settled on his nose. He ignored it. Charles pressed the camera button again and filmed the butterfly as it flapped into the air, circled and settled again on the dragon's nose.

'This,' I muttered a little louder, 'seems a bit silly. Doesn't the brute understand what we've built the hide for?'

Sabran laughed quietly.

'Is very OK, tuan.'

The smell of the bait drifted over to us and it occurred to me that we were sitting in a direct line between the dragon and the bait which had attracted him here.

Just then I heard a noise from the riverbed. I looked behind me and saw a young dragon waddling along the sand towards the bait. It was only about three feet in length and had much brighter markings than the monster close to us. Its tail was banded with dark rings and its forelegs and shoulders were spotted with flecks of dull orange. It walked briskly with a peculiar reptilian gait, twisting its spine sideways and wriggling its hips, savouring the smell of the bait with its long yellow tongue.

Charles tugged at my sleeve, and without speaking pointed up the streambed to our left. Another enormous lizard was advancing towards the bait. It looked even bigger than the one behind us. We were surrounded by these wonderful creatures.

The dragon behind us recalled our attention by emitting another deep sigh. He flexed his splayed legs and heaved his body off the ground. He took a few steps forward, turned and slowly stalked round us. We followed him with our eyes. He approached the bank and slithered down it. Charles followed him round with the camera until he was able to swing it back into its original position.

The tension snapped and we all dissolved into smothered delighted laughter.

All three reptiles were now feeding in front of us. Savagely they tore at the goat's flesh. The biggest beast seized one of the goat's legs in his jaws. He was so large that I had to remind myself that what he was treating as a single mouthful was in fact the complete leg of a full-grown goat. Bracing his feet far apart, he began ripping at the carcass with powerful backward jerks of his entire body. If the bait had not been securely tied to the stake, I was sure he would with ease have dragged the entire carcass away to the forest. Charles filmed feverishly, and soon had exhausted all the film in his magazines.

'What about some still photographs?' he whispered.

This was my responsibility, but my camera had not the powerful lenses of the cine camera and I should have to get much closer if I were to obtain good photographs. To do this would risk frightening the beasts. On the other hand, none of the giant lizards would be tempted by the small bait in the trap as long as the carcasses were within their reach, so if we were to capture one, we should have to retrieve the main bait somehow and re-hang it in the tree. It seemed that to try to take their photograph was as good a way of frightening them as any.

Slowly, I straightened up behind the hide and stepped out beside it. I took two cautious steps forward and took a photograph. The dragons continued feeding without so much as a glance in my direction. I took another step forward and another photograph. Soon I had exposed all the film in my camera and was standing nonplussed in the middle of the open riverbed within two yards of the monsters. There was nothing else to do but to go back to the hide and reload. Though the dragons seemed preoccupied with their meal, I did not risk turning my back on them as I returned slowly to the hide.

With a new film in my camera I advanced more boldly and did not begin photographing until I was within six feet of them. I inched closer and closer. Eventually, I was standing with my feet touching the forelegs of the goat carcass. I reached inside my pocket and took out a supplementary portrait lens for my camera. The big dragon three feet away withdrew his head from inside the goat's ribs with a piece of flesh in his mouth. He straightened up, and with a few convulsive snappings of his jaws, he gulped it down. He remained in this position for a few seconds looking squarely at the camera. I knelt and took his photograph. Then he once more lowered his head and began wrenching off another mouthful.

I retreated to Charles and Sabran for a consultation. Obviously, a close approach would not frighten the creatures away. We decided to try noise. The three of us stood up and shouted. The dragons ignored us totally. Only when we rushed together from the hide towards them did they interrupt their meal. The two big ones turned and lumbered up the bank and off into the bush. The little one, however, scuttled straight down the riverbed. I chased after it, running as fast as I could, in an attempt to

The dragon with the bait

catch it with my hands. It outpaced me, and as it came to a dip in the bank it raced up and disappeared into the undergrowth.

I returned panting and helped Charles and Sabran to hoist the carcasses into a tree twenty yards away from the trap and then once more we waited. I was fearful that having frightened the dragons once they would not return. But I need not have worried; within ten minutes the big one reappeared on the bank opposite us. He thrust his head through the bush and froze immobile. After a few minutes he came to life and descended the bank. For some time he snuffled around the patch of sand where the bait had been lying, protruding his great tongue and tasting the last remnants of the smell in the air. He seemed mystified. He cast around, his head in the air, seeking the meal of which he had been robbed. Then he set off ponderously along the riverbed, but to our dismay walked straight past our trap towards the suspended bait. As he approached it, we realized that we had not tied it sufficiently high, for the great creature reared up on its hindlegs, using its enormous muscular tail as a counterbalance, and with a downward sweep of his foreleg snatched down a tangle of the goat's innards. He wolfed it immediately, but the end of a long rope of intestine hung down from the angle of his jaw. This displeased him and for a few minutes he tried to paw it off, but without success.

He lumbered along the streambed, back towards the trap, shaking his head angrily. As he reached a large boulder, he stopped, rasped his scaly cheek against it and at last wiped his jaws clean. Now he was near the trap. The smell of the bait inside filtered into his nostrils and he turned aside from his path to investigate. Sensing accurately the direction from which the smell came, he moved directly to the closed end of the trap and with savage impatient swipes of his forelegs he ripped aside the palm-leaf shroud, exposing the wooden bars. He forced his blunt snout between two of the poles and heaved with his powerful neck. To our relief, the lianas binding the bars together held firm. Baulked, he at last approached the door. With maddening caution, he looked inside. He took three steps forward. All we could see of him was his hindlegs and his enormous tail. For an interminable time he made no movement. At last he went further inside and disappeared entirely from view. Suddenly there was a click, the trigger rope flew loose and the gate thudded down, burying its sharpened stakes deep into the sand.

Exultantly we ran forward. We grabbed boulders and piled them against the trap door. The dragon peered at us superciliously, flicking his forked tongue through the bars. We could hardly believe that we had achieved the objective of our four months' trip, that in spite of all our difficulties we had at last succeeded in catching a specimen of the largest lizard in the world. We sat on the sand looking at our prize and smiling breathlessly at one another. Charles and I had many reasons for

The captured dragon

feeling triumphant, but Sabran, whom we had only known for two short months, was as happy as we were and not only, I am sure, at catching the dragon, but at seeing us so jubilant.

He put his arm round my shoulders and smiled his broadest white smile. 'Tuan,' he said, 'is very *very* OK.'

20

Postscript

───

There is little more to be said, for we had achieved at last the objective of our expedition, and its aftermath was merely the inevitable struggle with bureaucracy necessary to export ourselves, our film, our cameras and our animals from Indonesia. The battle, though lengthy and hard-fought, was not always an acrimonious one and often it seemed that we and the officials concerned were fighting together against an overwhelming weight of restrictions and regulations which threatened to overcome us all.

I remember with affection the police officer in Sumbawa, the port to which we had sailed when we left Komodo. We were marooned there for a few days as the air service had been dislocated by an engine failure in Bali. There was no *losmen* in the town with a spare bed, so we slept on the floor of the airport building.

We had, of course, to visit the police station and we were received by a particularly charming man whose job it was to check our passports. He started with Charles's and began reading the copperplate writing on the inside cover which commences, 'Her Britannic Majesty's Principal Secretary of State for Foreign Affairs requests and requires . . .' Having digested this, he methodically pored over every visa and endorsement, making pencil notes on a grubby piece of paper, until at last he reached the list of foreign exchange with which Charles had been issued. This took a very long time, but as we had a four days' wait ahead of us, we were in no hurry. The police officer provided coffee and we provided cigarettes. At last he turned the final leaf of the passport and handed it back to Charles with the words, 'You are American, no?'

On our second day in Sumbawa, we were strolling past the police station when a sentry with a fixed bayonet sprang out at us. The officer required to see us again.

'Tuan,' he said, 'I am sorry, but I did not notice if you have in your passport a visa to enter Indonesia.'

On our third day, the officer came down to the airport to see us.

'Peace on the morning,' he said cheerily. 'I am sorry but I must see your passports again.'

'I do hope there is no trouble,' I said.

'No, no, tuan. But I do not know your names.'

He did not visit us on the fourth day so presumably his files were finally complete.

Our negotiations elsewhere, however, were not so satisfactory, and in the end we were refused permission to export the dragon. This was a great and unexpected blow to us, but we were allowed to take the rest of our animals – Charlie the orangutan, Benjamin the bear, the pythons, the civets, the parrots and our other birds and reptiles – back to London.

In one way I was not sorry that we had to leave the dragon behind. He would, I am sure, have been happy and healthy in the large heated enclosures of London's Reptile House, but he could never have appeared to anyone else as he did to us that day on Komodo when we turned round to see him a few feet away, majestic and magnificent in his own forest.

BOOK THREE

Zoo Quest in Paraguay

PARAGUAY

Miles
50   0   50   100

Marsh

BOLIVIA

BRAZIL

Chaco

Olimpo

Río Paraguay

Puerto Sastre

Puerto Casado

Puerto Pinasco

Pedro Juan

R. Verde

Tropic of Capricorn

Horqueta

Río Pilcomayo

Concepción

R. Monte Lindo

Estancia Elsita

San Pedro

R. Jejui

Puerto J.

R. Curuguati

Alto Paraná

A R G E N T I N A

R. Confuso

Rosario

Yhú

Asunción

Paraguarí

Villarrica

Río Paraguay

Pilar

Corrientes

SOUTH AMERICA

Paraguay

Río Paraná

Estancia Ita Caabo

↓ To Buenos Aires

21

To Paraguay

———

In 1958 we went to Paraguay to look for armadillos. To go so far in search of a creature whose attractions are not, perhaps, immediately obvious may need a little justification. There are many reasons why animals appeal to us. The exquisite beauty of birds, the grace and sleek strength of great cats, the dramatic, slightly horrifying appearance of giant snakes, the flattering affection of dogs, the mischievous, near-human intelligence of monkeys – all these are characteristics which win many devotees. But armadillos possess none of them. They are drably coloured and, except to the most sympathetic eye, not particularly beautiful. As far as I know, they cannot be trained to perform amusing tricks (to be truthful, I suspect that they are really rather unintelligent) and they do not make endearing pets. Yet they have one quality which, for me, is the most potently fascinating of any that an animal may possess – a blend of the exotic, the fantastic, and the antique which is only inadequately summarized by the word 'strange'.

It is not easy to define this characteristic. A lion does not possess it for a lion is, after all, essentially a larger version of our familiar domestic cat. Neither does a polar bear appear surprising or odd to us; it is after all very like a dog, merely somewhat larger, with a white coat to make it inconspicuous among the snows of the Arctic. Even such a curious creature as the giraffe is built on a pattern which is familiar to us, for it belongs to the tribe which includes the common European deer.

But there is nothing in Europe which remotely resembles

such extraordinary creatures as the kangaroo, the giant anteater, the sloth – or the armadillo. Both in their general appearance and their internal anatomy, they are quite unlike anything which inhabits our own continent; they are the last of their kind, the survivors from past geological ages when most of the animals of today had not yet appeared on earth. They, indeed, are truly 'strange'.

The causes of their survival are, in themselves, fascinating.

The kangaroo's ancestors once inhabited much of the surface of the earth. In their time, their newly developed ability to nurture their tiny embryonic young inside a pouch made them the most advanced creatures of the age. But as even more highly developed animals evolved – the placental mammals which were able to retain their young in a womb inside their bodies – the marsupials became outmoded, and were no longer able to win the battle for food and living space. As a result, the majority of them died out. Some – the opossums – survived in South America, but most of the marsupials of today are found in Australia, for that continent became separated from the rest of the world by an arm of the sea before the newer mammals evolved. Consequently, the old-style marsupial was protected from outside competition and lived on, in many different forms, until the present day. Australia, in fact, is a museum of living antiques.

South America, by virtue of its rather more complicated geological history, contains other strange animal antiques as well as the opossum. For many millions of years it was connected to North America by a wide bridge of dry land, but soon after the appearance of the first placental mammals it, too, became separated from the rest of the world. At this time the Edentates – the group which contains the sloths, the armadillos and the anteaters – were in the ascendant. During their period of isolation in South America, they evolved to produce a multitude of most extraordinary creatures. Giant sloths, almost the size of elephants, browsed in the forests. Glyptodons, relatives of the armadillo, possessing enormous bony shells, some over twelve feet long and armed with immense tails bearing great spikes at the end like medieval battleaxes, lumbered over the savannahs.

When the land connection with North America was re-established, about sixteen million years later, some of these fantastic beasts migrated northwards. They left their bones among the glacial deposits of North America, they fell into the pitch lakes of California, and they impressed their footprints on the shores of a lake in a country which was to become Nevada – footprints which were revealed at the end of the last century when workmen began quarrying sandstone to build a new gaol for Carson City.

The armadillos are the only surviving relations of the Glyptodons. To look at them is to see a link with the strange, primitive beasts of prehistory, and it is this more than anything else which makes them, for me, so intriguing. They live in burrows and trot through the forests and over the pampas, eating roots, small insects, and carrion. It may well be that they owe their continued existence in some measure to their armour-plated shells. Indeed, they seem to be an extremely successful group of creatures, for there are many different species of them, ranging in size from the tiny pygmy armadillo, no bigger than a mouse, which burrows in the sand of Argentina, to the giant armadillo that grows to four or five feet long and which roams through the hot moist forests of the Amazon basin.

While Charles Lagus and I had filmed and captured sloths and anteaters in Guyana neither of us had ever seen a wild armadillo. We hoped that we should do so in Paraguay. We also planned to look for many other birds, mammals and reptiles, but when the Paraguayans asked us why we had come to their country, I simply replied, without amplification, 'We are here to look for *tatu.*'

Tatu, I believed, meant armadillo. It is not Spanish but Guarani, the language which, with Spanish, is the official tongue of Paraguay.

My reply always had the same effect – roars of laughter. During our first days I had supposed that for some reason a man looking for armadillos must seem to be an irresistibly comic figure to all Paraguayans, but I began to mistrust this explanation as being too simple. When my reply reduced a senior official of the Banco Nacional del Paraguay to mirth bordering on hysterics, I felt the time had come to solve the mystery. Before I could say anything further, he asked me another question.

'What kind of *tatu*?'

I knew the answer to that as well.

'Black *tatu*, hairy *tatu*, orange *tatu*, giant *tatu*; all the different kinds of *tatu* that are found in Paraguay.'

He seemed to find this even funnier than my first reply. He was convulsed. I waited patiently for him to recover. Until then he had impressed me as a friendly and helpful man; he spoke perfect English and our conversation together had been most useful. His laughter subsided.

'Perhaps you mean some sort of animal?'

I nodded.

'You see,' he explained, '*tatu*, in Guarani, is also used as a not very polite word which means . . . well . . .' he hesitated, 'a sort of young lady.'

Why the armadillo, fascinating though it is, should have given its name to such a different being was not clear to me, but at last I understood the joke. In the months that followed, we were asked the same question many more times, but now I was able to give my reply with the conscious air of making a joke, and to use it, as a witticism, to soften the barriers of formality with ranchers, Customs officials, peasants and Amerindians.

On a few occasions, however, the joke misfired, for one or two people considered it quite unremarkable that we were wandering in the remoter parts of Paraguay in search of young ladies and were totally incredulous when we insisted that we really were looking for the four-legged variety of *tatu*. Sometimes, after we had cracked our *tatu* joke, some of our questioners went on to ask exactly why we were so interested in armadillos. I never managed to make myself clear on this point. My Guarani dictionary did not include a word for Glyptodon. Perhaps it was fortunate that it did not. I blench to think what, if it existed, its other more colloquial meaning might have been.

22

The Decline of a Luxury Cruise

At times, during previous expeditions Charles and I had been decidedly uncomfortable, and we often kept our thoughts from our aching legs or empty stomachs by trying to devise the ideal expedition on which we should live in lazy luxury, yet find the most beautiful and exciting animals in the world.

In New Guinea, we had walked several hundred exhausting miles to find some elusive birds of paradise. Towards the end of that journey, Charles had stated categorically that the first essential of his ideal expedition was some form of mechanical transport. On our voyage to Komodo when we had nothing to eat but salt fish and rice, I had stipulated that my priority would be an immense and infinitely varied larder of tinned delicacies. In a particularly flimsy camp in Borneo, we had both agreed, as we struggled to save our film and cameras from being soaked by torrential rains, that a completely waterproof habitation was also vital. In moments of less serious crisis but equal exasperation we had calmed our ruffled tempers by adding other details: I was determined to have an inexhaustible supply of chocolate; Charles to sleep somewhere that was infallibly proof against beetles, cockroaches, ants, centipedes, wasps, mosquitoes and all other biting or stinging insects. This hypothetical expedition eventually became very real to both of us, but neither of us imagined that it would ever be translated into fact. Within a week of our arrival in Paraguay, however, a British meat firm in Asunción spontaneously offered us the means of making a journey which seemed to match these specifications almost exactly.

This embodiment of our dreams was called the *Cassel*. She was a thirty-foot-long, diesel-powered, capacious cabin cruiser with a draught so shallow that she should be able to take us, our equipment and our food up the small, tortuous rivers of the far interior without the slightest difficulty on either her part or on ours. We accepted the offer of her loan with alacrity and enormous gratitude.

As we steamed past the wharves of Asunción, up the wide brown Rio Paraguay, we stowed our cameras and recording equipment into roomy bone-dry cupboards in the cabin. We piled the galley high with packet soups, sauces, chocolate, jars of jam and tinned meats and fruit in amazing variety. We fastened double layers of mosquito netting over the window frames. I arranged above my bunk a small library of paperback books. Charles tuned the little radio to a station in Asunción and filled the cabin with haunting guitar music.

More than satisfied with our luxurious accommodation, I walked aft to have a loving look at the dinghy that the *Cassel* was towing from her stern. It was fitted with a 35-horsepower outboard engine and we already referred to it, reverently, as the speedboat. In it, we hoped to roam far and wide among the smaller tributaries looking for animals and returning to the *Cassel* for food and sleep.

There was nothing further to do. I climbed on to my bunk and relaxed. We were on our way, in unprecedented comfort, to the southern fringes of the great tropical forest which begins in the north-eastern part of Paraguay and extends across Brazil to the Amazon Basin and beyond towards the Orinoco – the largest area of primary jungle in the whole world. It seemed too good to be true.

It was. Within ten days we were to be more wretchedly uncomfortable than we had ever been on any previous expedition.

We had three companions on board. Our guide and interpreter was a husky, brown-haired Paraguayan who spoke fluent Spanish, Guarani, and one or two Indian languages. In addition – and astonishingly, for he had never been outside South America in his life – he spoke English with a broad Australian accent. His name was Sandy Wood.

Paraguay is full of people of foreign stock. Poles, Swedes, Germans, Bulgarians and Japanese, all have flocked to this small republic in an effort to escape land shortage or religious oppression, political tyranny or the law. Sandy's parents had come with nearly two hundred and fifty other Australians just before the end of the last century. At that time there had been a disastrous general strike in Australia and a journalist named William Lane, who had long preached an ideal form of socialism, banded together a group of farmers, carpenters and other working people who shared his views and brought them to Paraguay to set up his perfect society. The Paraguayan Government gave the immigrants good arable land and the new community of Nova Australia was founded. All property was to be communally owned; those who joined the group had to give all their money and possessions to the colony's exchequer; everyone had to work, not for individual wages but for the common good. These high-minded political ideals were combined with a dash of puritanism. There was to be no contact with the local people, no drinking of spirituous liquors, no music, no dancing.

Within a year, the strain of living in conformity with such high principles began to tell on the community. The attractions of the pretty Paraguayan girls, the taste of *caña*, locally fermented cane-juice, and the gaiety of the guitar-playing villagers caused some of the colonists to backslide. Even more serious for the young community's economy, some of the less energetic colonists began to leave the hard work to others and to content themselves – in Sandy's expressive phrase – with doing the grunting.

Colonia Australia was a failure. Undiscouraged, Lane founded another colony, Colonia Cosme, on a new site and took with him the few people who had remained true to their original principles as well as some new immigrants from Australia. But this venture, too, was unsuccessful. Its members began to defect. A Paraguayan revolution was fought over the colony's land and the buildings were sacked first by the revolutionaries and then by the avenging Government forces. The members of the community scattered. Many went down to Buenos Aires to work in the railyards. Some went as far afield as Africa to try farming. A few remained in Paraguay and earned their living as loggers, farmers and carpenters. Sandy's parents had been among them, and Sandy himself had been born and reared a Paraguayan. He had tried a variety of jobs. He had felled timber in the higher reaches of the river up which we were planning to travel, he had worked cattle on an *estancia*, he had been a hunter, and currently he held a spasmodic and ill-defined job in a tourist agency in Asunción. His linguistic skills, his knowledge of the forest and his placid temperament made him an ideal guide for us.

The other two people on board constituted the official crew. There was some doubt as to which of them was actually captain. Gonzales, the thinner, taller and more cheerful of the two, wore a nautical cap. Originally it had been handsomely wreathed with gold braid but it was now so battered that the braid had come adrift and hung drunkenly down the peak. This cap, he told us in confidence, was really the insignia of captain, but in addition to possessing all the skills that this exacting post required, he was also uniquely qualified to tend the engine. However, he had decided that even he could not deal properly with both jobs at once, and in consequence he was willing to allow his companion to be referred to as 'Capitan' even though, as he was at pains to explain, this was in reality only a courtesy title.

Capitan was short and alarmingly pot-bellied. He habitually wore an enormous bell-shaped straw hat with its brim turned down and, beneath it, dark glasses. He did not remove these glasses even in the evenings so that we were tempted to speculate as to whether or not he wore them in bed. His mouth was set in a down-pointing semi-circle of permanent pessimism and his cheeks were marked by some kind of skin complaint which produced un-sunburnt patches of livid pink. He occupied his leisure moments – and he seemed to have many of them – by smearing these patches with a special ointment. His unfailing rejoinder to any comment, question or observation, was to suck his teeth dismally.

To reach the remote area of forest in which we had decided to work, we had to travel north up the Rio Paraguay for some seventy-five miles and then turn eastwards along one of the Paraguay's great tributaries, the Rio Jejui. We hoped to continue along it until we reached its lonely headwaters where no one lived except Amerindians and a few timber men. The journey would take us at least a week.

During the first few days, we spent much of our time lying on deck watching the *Cassel*'s bows knifing the brown waters of the river and severing the rafts of *camelote*, floating water hyacinth whose elegant spatulate leaves, swollen at their base into buoyant air bladders, shrouded clusters of delicate mauve flowers. The larger clumps we avoided, for even if our boat could have cut into them, their matted dangling roots would have fouled the propeller. Some of these islands carried their own passengers – herons, egrets, and, prettiest of all, chestnut coloured lily-trotters which stepped fastidiously among the camelote leaves, lifting their long-toed feet high as they searched for the little fish which had misguidedly taken shelter there. As we approached, they took fright at the noise of our engine and flew up, flirting their yellow underwings. They circled us, their long legs dangling limply, and then landed again on their camelote raft which was by now behind us, undulating in our wake.

Sandy sat in the stern sipping *maté*, Paraguayan tea. Gonzales squatted by his engine, enthusiastically strumming a guitar and singing loudly, though none could hear him for the engine completely drowned the sound of his voice. Capitan was perched on a tall stool in the wheelhouse, steering the ship with one

Rafting timber on the Rio Jejui

hand and smearing ointment on his face with the other. It was fiercely hot. In an endeavour to keep cool, Charles and I went below and lay on our bunks; but there it seemed even hotter, for in the shelter of the cabin there was no breeze to evaporate our sweat, and soon our sheets were drenched.

Suddenly the engine stopped and the unaccustomed silence was filled with the shrill voices of Gonzales and Capitan arguing fiercely. We scrambled on deck. Sailing sedately downstream behind us, we saw two cushions and a seat. The speedboat itself had disappeared entirely. Sandy explained unemotionally that Capitan had just put the boat into a sharp turn in order to avoid a clump of camelote and the speedboat had capsized in our wake. Gonzales was leaning over the stern ineffectively tugging at the speedboat's painter which was still attached to the bollard but which descended almost vertically into the muddy water. Capitan sat in the wheelhouse with an outraged expression on his face, vociferously sucking his teeth.

In the argument that followed, it transpired that neither Capitan nor Gonzales could swim. While the pair of them continued their recriminations Charles and I stripped off our clothes and clambered overboard. Fortunately and unexpectedly the river here was not deep, but even so it took us nearly two hours to drag the submerged speedboat to the shallows, to right it, and to recover the engine, the three fuel tanks and the toolkit. By the time we had finished, the seat and the cushions must have been well on their way to Asunción, for though we returned downstream for a mile or so we did not find them. It was not a real disaster, but it made us suspect that Capitan's watermanship was not impeccable.

The next three days were uneventful. We left the Rio Paraguay and turned eastwards up the Jejui. A few miles up it we stopped for an hour or so at Puerto-i, the last village of any size on the river. When we left it, Capitan's habitual expression of gloom deepened perceptibly. He had not been along the Jejui before, and now that he was doing so he did not like it. These were dangerous waters and he sensed an impending catastrophe. He faced it on the morning of the fourth day. Ahead of us the river twisted into a hairpin bend round which its waters rushed in a series of blistering whirlpools.

Capitan stopped the engine with an air of finality. He had already, against his better judgement, performed miracles of navigation, but the hazard ahead was totally unnegotiable. We must go back. After persuasion, he consented to inspect the bend in the speedboat. When he returned, his expression made it clear that this closer view had only confirmed his worst fears.

There then developed a somewhat emasculated argument. Sandy, as interpreter, seemed to have decided that it was better to convey to each party only those sections of the discussion which were strictly germane to the problem in hand and to omit the scathing personal references which I suspected Capitan was making about us – and which we were certainly making about him. To us, the bend seemed tricky but by no means impossible. To slink back to Asunción, having wasted at least a week, was unthinkable. Neither could we work in the surrounding country for it was semi-cultivated. We should never find in it the animals we were seeking. But Capitan's mind was made up. He did not wish to die, he said somewhat melodramatically, and he did not suppose that we wished to do so either. We retorted scornfully. Sandy made a polite translation. As we argued, a filthy little launch, its engine knocking painfully, crawled past us and unconcernedly disappeared round the bend.

As we watched it, our fury redoubled. Not to be able to convey the full strength of our wrath directly to Capitan, without Sandy as middleman, became almost unendurable. We all lost our tempers. It was Charles who finally produced the two words of Spanish which, for the first time, enabled us to establish direct and unambiguous communication with Capitan. An hour or so earlier, Capitan had been struggling with a paraffin stove and he had explained that it never worked properly and was always giving trouble because it was not of European manufacture but 'Industria Argentina'. Now, Charles pointed vigorously at Capitan and said with all the venom he could muster, 'Capitan – industria Argentina'. He seemed so pleased with himself at this linguistic triumph that we all laughed. Sandy seized the moment to retreat tactfully and make himself some more maté. Without him, the argument necessarily came to a halt.

Charles and I joined Sandy to discuss the situation. We recalled the little boat which had passed us during the heat of the argument. If one was going upstream, there might be another which might be able to give us a lift. Consoling ourselves with this last shred of hope, we ate our evening meal and went to bed.

We were woken at midnight by the sound of a launch. We rushed on deck and shouted as loud as we could and it drew alongside. Fortunately Sandy knew its captain, a small, swarthy man named Cayo, for the two had met when Sandy had been logging in this area several years before. For ten minutes, he and Cayo talked, flashing their torches over the launch and its cargo, over the *Cassel* and on to one another's faces. Charles and I waited patiently in the darkness outside the torch-lit arena of discussion.

At last Sandy turned to us. Cayo was on his way to a small logging settlement high up in the head waters of the Rio Curuguati, a tributary of the Jejui. This was exactly the area we had hoped to reach. However, Cayo already had three passengers – axemen who were going to work at the camp – and a full cargo of stores. There was no room for us, but he could take the essential part of our equipment and a minimum quantity of food. We ourselves could make the journey in the speedboat.

'How do we get back?' I murmured, almost ashamed of myself for asking such a chicken-hearted question.

'Well, that's a little uncertain,' said Sandy airily. 'If the river is high, Cayo may spend a few days up there looking around. If it's not, he'll come back immediately and we might be stuck for three or four weeks or more.'

There was no time for a long discussion. Cayo wanted to be on his way. We decided to take the risk of being stranded, sealed the bargain with an advance payment and rapidly transferred our equipment on to Cayo's launch.

Half an hour later, Cayo left us, taking with him several thousand pounds' worth of cameras and recording gear. We watched the yellow stern light grow smaller in the blackness of the night and then finally disappear as he rounded the bend. Charles and I went back to our bunks assuring one another that neither of us was in the least apprehensive about our new plan of action. Supposing we are stuck for a week or two, we said, it will be rather fun, won't it? Neither of us seemed very sure.

The next morning we said goodbye, with slightly forced affability, to Capitan and Gonzales, untied the speedboat and roared away upriver in pursuit of our equipment. I did not have time to snatch a last look at the *Cassel*, for Capitan had had good cause to be baulked by the bend. Its eddies and whirlpools snatched alarmingly at the hull of the speedboat, and we skated and slid over the surface of the water in a most frightening fashion. By the time we reached a smoother stretch of the river, the *Cassel* was out of sight. I was sorry. I would have liked one last lingering look at the vessel which still contained in its waterproof mosquito-proof cabin, our luxury foods, our little library, the radio and the comfortable bunks. It was sad to have deserted our ideal trip almost before we had started. As we raced on up the river, the forest on the banks seemed depressingly desolate and threatening. Heavy storm clouds were gathering in the sky ahead of us.

I experienced a nasty sensation of nakedness.

We soon caught up with Cayo although he had been travelling continuously since he left us. His launch was labouring valiantly onwards but she was loaded to the gunwales and was not making more than three knots, while we in our speedboat had been going at six times this speed. The prudent course would have been to tie up behind the launch and remain close to our cameras, food and bedding, but there was no room for us on board and to have tied our speedboat behind would have slowed down the launch even further. We decided to abandon uncomfortable caution and continue our exhilarating surge up the river, taking with us a camera in case we saw something filmable, our hammocks and enough food for three meals in case we failed to meet the launch again that night.

Cheerfully we swept round the bends of the increasingly serpentine river, our gigantic wake fanning out from our stern to roll in waves along the river banks and lose itself among the bushes and creepers which grew close to the river's edge.

As we went farther upstream the size of the forest trees increased until soon we were speeding between high green walls beyond which rose the rounded domes of giant hardwoods – *quebracho* and *lapacho*, *curupay* and Spanish cedar – the valuable prizes which tempt men into this deserted country. Our roaring engine flushed birds from the banks – nose-heavy toucans, scarlet macaws always in pairs, flocks of parrots and, most commonly, black hangnests with scarlet rumps which shot shrieking from their club-shaped nests that dangled in groups from the branches of trees overhanging the river.

Once again we were alone with the forest; and once again it seemed brooding and malevolent. As we raced past its lowering green walls, the white water sparkling and tumbling from our stern and glistening in the sunshine, we seemed so close to it and yet in such a different world, that I felt the same thrill as comes from sitting comfortably indoors while outside, the thickness of a pane of glass away, it is cold, wet and unpleasant. But I knew that if the engine were to fail, if we were to strike a submerged log and rip a hole in the bottom of our boat, if the looming blue storm clouds on the horizon were to break into a heavy rain-storm, we might be faced with a prospect which would certainly be extremely uncomfortable and might well be disastrous. I thought longingly of the comfort and security of the *Cassel's* cabin.

At sundown, we reached the mouth of the Curuguati. We decided to camp on the bank and wait for Cayo. It was a wretched camping site. The neck of land between the two rivers had been cleared of bush and a squalid shanty built by the loggers who sometimes used the place as a base from which to travel into the forest and fell trees. The clearing was littered with rusting wire, empty oil drums which would be used as floats for the rafts of heavy timber, and patches of diesel oil, spilled by the launchmen when refuelling their boats. It was deserted except for an Amerindian boy who lounged by the cabin morosely watching us as we tried to improvise supports for our hammocks among the rusty drums.

During the night we heard Cayo's launch stolidly chugging towards us. He did not stop. We merely exchanged shouts with him as he turned into the Curuguati, promising that we would catch him up again the next morning. Then we went back to sleep.

When dawn came, we lost no time in packing up and starting off again.

We took turns at the wheel of the speedboat. Sandy drove at a breakneck speed which I found frightening. With his hat pulled down tight on his head, its front brim blown vertical by the wind, he sat imperturbably spinning the wheel as we careered round bend after bend, the boat heeling over until we almost shipped water, its stern skidding wildly over the surface of the river. I lay back in the stern, dizzy with apprehension, and shut my eyes.

Suddenly Sandy shouted a warning yell. There was a fearsome crashing of branches, a hideous scraping noise, a jolt which threw me out of my seat, and we came to a halt. Our bows were halfway into the bank, and the boat was rocking alarmingly in our own wake which had caught up with us. The steering cable had broken just as Sandy had spun the wheel to whip us round a sharp bend.

There was only room for one of us to tackle the repair work, and I undertook it. Without pliers and a steel spike it was not possible to splice the frayed cable; the only hope was to knot and bind it. I worked as swiftly as I could, but it was a very unpleasant job and to refasten the wire to the steering column I had to lie with my head in the covered forepeak of the boat. The heat was so intense that I was soon cascading sweat. The wire strands of the cable pricked and cut my hand and I became covered in oil and grease. To increase our discomfort, we had stopped in a place that was haunted by swarms of vicious mosquitoes which bit us maddeningly and continuously. All the time I was thinking of Cayo, steaming steadily away from us, taking with him our only supply of food and our equipment. This was the sort of disaster that I had been dreading.

An hour passed before we were able to start again. In a more sober fashion we continued up the river. To my surprise, the improvised repair seemed to hold quite well, though there was continual danger that the knot might jam as it wound round the steering column.

At last we caught up with Cayo and once again we passed him. I breathed a sigh of relief. Now if the steering broke down irreparably, all we should have to do was to wait for him to catch up with us.

In the early afternoon the heavy sullen clouds which had been ominously building up for the past few days, burst with a loud clap of thunder. Heavy drops of rain began to pockmark the surface of the river. Then the engine stalled. Despairingly we tugged at the starting rope. It coughed into action just as the worst of the storm arrived.

The remaining hours of that day were miserable. The rain fell so heavily that the view ahead of us was blotted out as though by thick mist. The engine began to stall with increasing frequency, but we dare not remove the cowling to trace the cause of the trouble in case the drenching rain reached the plugs and the carburettor and stopped the engine for good. It was wretchedly cold. Sandy drove doggedly on. I sat beside him, keeping watch over the knotted cable. Charles lay in the stern ready to pull the starting rope when the engine failed. When it was running he covered himself with an old leaky tarpaulin in a more or less successful attempt to keep dry and warm. At the beginning of this journey he had decided to grow a beard and he had also equipped himself with an American-style baseball cap with a long peak. Neither Sandy nor I thought it suited him. Now, whenever we stopped, this curious bearded, capped figure peered out of the tarpaulin, smoking a cigarette in a long holder, rain water pouring down his face and dripping off the end of his nose, to swear eloquently and with great deliberation.

We drove on through the storm. The camera and film had been stowed in the forepeak where we hoped they would remain dry. Sandy claimed that we were now quite close to the hut occupied by a timber-man and his wife, which was our final destination. Each time we rounded a bend, I searched hopefully for it. The engine sputtered to a halt again and again and remained obstinately silent until Charles beat it into action with savage pulls on the starting rope. Twice the steering cable parted and needed re-securing. The sun had disappeared in the cloud-soaked sky many hours before, but the darkening river told us that it must have set and that evening was on us. It was almost dark when we turned a bend and saw, far away up the new reach of the river, a pinpoint of yellow light. By the time we reached it, the night had long since fallen. We moored the boat at the foot of a small cliff and ran to the house up a steep, narrow path, which in the continuing storm had turned into a waterfall.

The light came from a large fire of logs burning in the centre of the earth floor of a small rectangular hut. There was no door. Crouching round the fire, their faces illuminated by the flames, squatted a young woman dressed in a long-sleeved blouse and trousers, a black-haired man of about thirty, and two Amerindian youths. The noise of the lashing rain and the screaming wind drowned the sound of our footsteps so that the occupants of the hut were unaware of our presence until we stood, dripping, on their threshold.

The man jumped up and welcomed us in Spanish. There was not time to make lengthy explanations for our baggage and equipment was still out in the storm and he ran with us back to the boat to fetch it.

After a meal of hot soup, our host showed us into a store-room where he said we might spend the night. It was full of barrels, bulging sacks, greased axes and pieces of rusting machinery, all draped with cobwebs. Huge brown cockroaches covered the mud walls in a glistening, moving carpet and, above, bats were flitting among the rafters. The room reeked with the nauseating stench of putrefying salt beef. But it was dry. Gratefully we slung our hammocks as the thunder crashed over the forest outside. Within a few minutes we were asleep.

23

Butterflies and Birds

The storm blew itself out during the night and in the morning the sky was cloudless and intensely blue. The settlement at which we had landed was called Ihrevu-qua, a Guarani name which meant 'Place of the Vultures'. Our host, Nennito, and his wife, Dolores, also owned a small modern house in the town of Rosario, but they seldom visited it for Nennito had been granted by the Government a concession to work timber in the forest up here on the Curuguati. If he could fell all the trees that in theory belonged to him, and float them down the river to the sawmills in Asunción, he would be a rich man. He did not do any of this work himself, however, for he was the *patron* whose function was a strictly supervisory one, and he engaged teams of men, like those coming up on Cayo's launch, to do the felling, hauling and rafting. When there were no men to supervise, as was the case when we arrived, he had nothing to do except sit outside his house and drink maté.

Although he had lived at Ihrevu-qua for several seasons, he seemed to have done little to make his home comfortable. There were no mosquito nets over the windows and there was little furniture. No banana or pawpaw trees had been planted. Dolores cooked on an open woodfire and she had no refrigerator. The rigours of this uncomfortable existence had already begun to show on Dolores's handsome fine-boned face.

Nonetheless, they were a happy, cheerful pair and most hospitable. Their home, they said, was ours for as long as we wished to stay.

286

There were several buildings in their little homestead, all of them interconnected by covered verandas: the kitchen, in which the fire was perpetually burning; the storehouse in which we had slept on our first night; a bedroom for Nennito and his wife; another for the two Amerindian boys and a third outhouse which was occupied only by chickens and a few stores until we used it later to sleep in. From the huts, the ground sloped away towards the river until it fell more rapidly in a steep incline of smooth red sandstone. At the foot of the rocks ran the turbulent brown waters of the Curuguati, swollen by yesterday's storm. Behind the huts Nennito had planted a small patch of cassava and maize, and beyond began the forest itself.

On that first morning, the whole of the clearing was filled with a blizzard of butterflies. It was an astonishing sight. So numerous were they that a single sweep of my net caught thirty or forty of them. They were beautiful creatures, their forewings iridescent blue, their hindwings scarlet and their undersides patterned with luminous yellow hieroglyphs. I recognized them as belonging to the genus Catagramma.

Butterflies are known to migrate in astonishing numbers and over vast distances. The great American zoologist, Beebe, once saw a swarm of migrants which flew over a pass in the Andes at a rate of at least a thousand a second and continued to do so in an unbroken stream for several days. Many other travellers and naturalists have noted the same thing. But the Catagrammas at Ihrevu-qua were not migrants, for they flew only in the clearing by the huts. A few yards away in the forest or further downriver there was none to be seen. We found that we could predict the appearance of these Catagramma swarms. They came always after a heavy storm when the sky was clear and the sun shone so fiercely that the rocks by the riverside were hot enough to be painful to the naked foot.

As night approached they gradually disappeared until by the

A blizzard of butterflies

time it was dark they had all gone. If the next day was not hot and sultry, then they did not appear at all. Perhaps this particular type of weather caused the chrysalises to hatch in their tens of thousands and so cause the swarm. But where did all these insects go to at nightfall? A butterfly's life is a short one but they could hardly all have died at the end of the day. Did they fly to the forest and there roost, rank upon rank, beneath the leaves of the taller trees? I did not know.

Catagrammas were not the only species of butterfly that flew around Ihrevu-qua. Nowhere else had I seen so many. Not only were there vast numbers of individuals but also a great number of different species. To amuse myself when we had nothing better to do, I began to collect some. I did not work industriously or go out of my way to look for them. I did not beat bushes and explore the swamps in the way that a true lepidopterist would have done. I merely tried to catch a specimen whenever I happened to notice a butterfly of a kind I had not seen before. Yet in the fortnight that we spent in and around Ihrevu-qua, I collected over ninety different species. There must have been at least twice as many in this small area if I had had the patience and skill to have caught them. How remarkable this number is can be judged from the fact that in the whole of Great Britain only sixty-five species have ever been found and these include the rarest migrants.

The most gorgeous and the largest of all the butterflies I saw lived only in the forest. This was a species of the magnificent Morpho group, and like nearly all its relations it was emblazoned on its upperside with a wonderful shining blue. It measured over four inches across its wingtips. At first, when I saw one flapping lazily and erratically through the forest, I would set off in pursuit, running through the undergrowth, my shirt catching in the thorns, trying to follow it as it twisted and turned, and flailing my net wildly. But the Morphos knew when they were being chased, or – to put it with more scientific exactitude – they behaved differently when they were alarmed, for once I had made a close but abortive swipe at them with my net they would change their gait immediately and fly rapidly and straight, often soaring high among the branches and well out of reach. It was only after I had made several of these useless, sweat-provoking sprints that I realized that I should have to change my tactics.

The Morphos seemed to prefer flying in clear spaces, unimpeded by branches or bushes, and they therefore favoured the wide paths that Nennito's men had cut in the forest so that they could easily haul the logs down to the river. The Morphos often floated down these rides, their wings flashing as they caught a shaft of sunlight. At first I would go towards them, my net raised for action, but if I were moving by the time they came close to me, they would take fright and swing suddenly away, up into the tangled forest. The better technique was to stand quite motionless, net poised, until the insect, unknowingly, was within reach, and then try to net it with one sweep. It was not unlike a game of cricket and certainly the tricky jerking flight of the Morphos was as deceiving to the eye and as unpredictable as any googly delivered by a Test bowler.

There was, however, a much easier way to catch them. Professional butterfly hunters tempt the insects down with a bait, usually a mixture of sugar and dung. But here it was unnecessary. The forest was full of wild, bitter orange trees and much of the fruit lay rotting on the ground. The Morphos came down, nearly always in pairs, to sip the fermenting juice. Even when they were feeding, however, I had to be cautious and stealthy in my approach and accurate in the pounce if I were to net them.

Other butterflies had different tastes. On one of my walks in the forest, I smelt a nauseating whiff of putrescence. I traced it to its source and found the decomposing body of a large lizard. But I had difficulty in recognizing it, for it was almost completely concealed beneath the quivering mass of butterflies whose wings were patterned with a subtly modulated design in shades of midnight blue. They were so absorbed in their repellent feast that I was able to pick them off by their closed wings with my thumb and forefinger.

Though the Catagrammas, the Morphos, and the rest of the forest butterflies were abundant, none of them could compare, in sheer numbers, with the brilliant hordes that congregated at the edge of the river.

My first view of one of these great assemblies took me totally by surprise. One day, I stepped out of the moist twilight of the forest into a sunlit meadow of lush grass studded with small palm trees. A little stream glided silently through sedge and moss from one deep brown pool to another. I stood quietly and still in the shade of the trees behind me, searching the meadow with my binoculars, for I was anxious to spot any creatures that might be grazing or fishing in the stream, before I made myself conspicuous by stepping out into the sun and so scared them away. It seemed deserted. And then I saw that on the far side the stream was smoking. For a moment I thought, quite illogically, that I had found a hot spring, or perhaps a sulphur fumarole like those that erupt on the flanks of quiescent volcanoes. But a second's thought told me that there could be no volcanic activity in this area. Puzzled, I walked towards the smoke. It was only when I was within fifty yards of it that I realized with certainty that I was looking at a cloud of butterflies of a concentration that I would not have believed possible.

As I came close to them, the ground seemed to explode silently into a vast yellow cloud, but even as I stood there, amazed, the butterflies settled to the ground again. They were so densely packed that, as they sat with their wings folded, their bodies almost touched and it was difficult to see the sand beneath them. A few yards away, on the fringe of this tremulous yellow carpet, a group of black ani birds were busily feasting on the unresisting butterflies. They were oblivious of the birds and of me.

Swallowtails drinking

Each one had uncurled its proboscis, which is normally carried curled like a watch spring beneath its head, and was feverishly probing the wet sand. They were drinking. Yet as fast as they drank, they squirted little jets of liquid from the tips of their abdomens. They could not be short of water. It seemed more likely that they were absorbing the mineral salts dissolved in the water as it passed through their bodies. I squatted down to watch them more closely and as I did so they confirmed my suspicion that they were seeking salts, for as soon as I had stopped moving they settled on my arms, my face and my neck. They found my sweat as attractive as the mineral salts of the swamp and soon there were several dozen on me and others were circling my head, their wings making a loud, dry rustling in the air. I sat still and felt their tiny, thread-like proboscies gently exploring my skin and their delicate legs almost imperceptibly pattering across the back of my neck.

Butterflies drinking sweat

Though this sight and this experience was to become familiar in the weeks that were to follow, it never lost its fascination. We found these drinking assemblies not only by streams and marshes but even more commonly on the silver beaches and sandspits of the river above Ihrevu-qua. There, on any sunny day, we could be certain of seeing these gaudy swarms. In addition to the yellow species that I had first found, there were many other kinds, each tending to settle in a group by itself. I counted over a dozen different kinds of swallowtails alone. They were large, handsome creatures and as they drank they always quivered their wings. Some were velvet black with carmine blotches on their wing tips, some were yellow with black bars and patches, some had almost transparent wings marked only by delicate black veining. The reason each kind assembled in a group by itself seemed to be that each butterfly was attracted to its own image. If one, flying by, caught sight of another coloured like itself, then it settled down by it until, in a few minutes, there was not one but forty or fifty similar insects. Yet they were not always identical. Their powers of vision may not have been perfect, for when I came to examine the groups in detail I often found that in each there were several kinds which, though they resembled one another superficially, were in fact quite distinct, differing sometimes in size as well as in the details of their patterning. At first I thought that these might be individual variations or perhaps sexual differences, but when later I came to identify them scientifically, I found that they were distinct species.

As we travelled up the river in our boat, the wash from our wake often rolled up in waves across these sand beaches. It crashed into the drinking butterflies and overwhelmed them. When the water receded, it left behind a sodden, bedraggled patch of broken wings and bodies on the sand. Yet even these still contained the colours and shapes which attracted the flying butterflies and within seconds more had come down to settle on the corpses.

Unfortunately, butterflies were not the only insects that were unusually abundant at Ihrevu-qua. We were tormented by hordes of vicious biting pests. They were not only the most ferocious that I had ever encountered but they also had another distinction. They worked on a rigid shift system.

At breakfast time, the mosquitoes were on duty. There were several sorts, the most savage being a large kind with a distinctive white head. We usually ate our breakfast close to the fire in the hope that its acrid smoke would keep them at bay, but some of them endured even that in order to suck our blood. By the time the sun had risen above the forest on the other side of the river and was baking the red earth of the clearing, turning it to dust, the mosquitoes had forsaken the house and retreated to the shade beneath some of the trees which overhung the river. They would still bite us as enthusiastically as ever if we were incautious enough to go down there, but as far as people in the house were concerned, they were off duty.

Their responsibilities were taken over by the *mbaragui*, large flies like bluebottles, whose bite felt like the stab of a needle and left a small spot of red blood beneath our skin. The mbaragui were hard-working. They pestered us unmercifully throughout the heat of the day, but when dusk approached, they retired. A few of the mosquitoes might now return to work, but the main responsibility for our persecution was borne by the *polverines*. These were tiny black flies no bigger than a speck of dust and they were perhaps the most unpleasant of all. At least the mosquitoes and the mbaragui were large enough to be caught, and when you smack one which has its proboscis deep in your skin and burst its distended abdomen so that the blood it contains spatters your skin, you feel some sense of satisfaction – even though it is your own blood. The polverines, however, were so small and numerous that even though we massacred fifty with a slap it seemed to make no difference to the hazy black cloud which hung around our heads. What is more, there was absolutely no protection against them. Our mosquito nets had quite a fine mesh, but the polverines passed through them without any trouble. The only material which was sufficiently closely woven to stop them was a normal sheet. We tried making a tent of one, but it was so hot and suffocating inside that we had to abandon the idea. Instead we plastered ourselves with citronella, and with several other patent insect repellents, some of which smelt disgusting and others which merely made our skins smart and our eyes and lips sting excruciatingly. The polverines seemed to regard these medicaments merely as piquant dressings to their meal. They feasted on us throughout the night. As dawn came they went off watch and the mosquitoes once again took up their duties.

The only variations in this schedule were produced by the weather. If the day was particularly muggy with a heavy, overcast sky, or if the night was brightly moonlit, then mosquitoes, mbaragui and polverines all worked simultaneously. There was only one kind of weather which drove them all away – heavy rain. The fact that at Ihrevu-qua at least one day in every four was wet would normally have filled us with despair for we could not film. But here these rainy days were happy ones of enormous relief. The heat was less intense and we lay in our hammocks, reading in a blessed state of freedom from insects.

During our first days, we had an additional and very serious worry. According to our calculations, Cayo should have arrived at the shack about twenty-four hours after us. But he had not. Soon the last of the tinned food we had brought with us in the speedboat was finished and we had to ask Nennito if he could supply us. We were reluctant to do so, for not only were we already in his debt for our accommodation, but also his diet, which we were asking to share, was neither varied nor palatable. It consisted only of boiled cassava and stale salt beef, with perhaps a few wild sour oranges. But he could not supply us with fuel for our speedboat and our tanks were almost empty. The situation was serious. If Cayo had broken down soon after we had last seen him and had stayed there, we should just have enough petrol to take us back to him. If, however, something irreparable had happened to his engine and he had decided to try to drift and pole his boat back to the Jejui, then he would be beyond our reach by engine and we, too, might be reduced to the lengthy and hungry course of drifting down after him.

As the days passed we became increasingly concerned. But on our fifth day at Ihrevu-qua he arrived smiling cheerfully, as though nothing had happened.

We threw him a rope and within a few seconds he had moored the launch and was striding up the rock slope to the huts. I stayed by the river long enough to see our box of tinned food put ashore and then followed him.

Sandy, Nennito and Cayo were sitting round the fire, drinking maté, which was being dutifully replenished and handed round by Dolores. Maté is the crushed, dried leaves of the yerba maté tree, a shrub related to the holly. It is put in a horn or a gourd; water, hot or cold, is poured on it; and then the infusion is sucked through a *bombilla*, a tube with a strainer on the end. It has a bittersweet, astringent taste and both Charles and I were beginning to be very fond of it. We joined them.

'Cayo had a bit of engine trouble,' Sandy told us, 'but it is all right now. He says the river is high, so he is going farther upriver to see what the timber is like up there. If the water stays high, he will be gone for a couple of weeks. If it begins to fall, he'll come back sooner and in a hurry. Either way, he will pick us up and take us back to Asunción.'

It seemed a good arrangement. Cayo put on his hat, shook hands all round and went back to his boat. In a few minutes he was out of sight.

Now that our return journey was assured we could give our minds to the collection and filming of animals. The first thing to do was to enlist some help. Several pairs of eyes and hands are better than only three, particularly if the additional ones belong to Amerindians who know the forest and its inhabitants better than any European. Nennito told us that there was a *tolderia*, an Amerindian village, some five miles away through the forest. Sandy and I set out to find it.

The tolderia proved to be a group of dilapidated thatched huts in a pleasant little valley, wide, unwooded and verdant. The Amerindians had largely abandoned their traditional way of life. They wore tattered European clothes, and instead of hunting for meat in the forest they kept a few scraggy chickens and some half-starved cattle whose ribs stuck painfully from their sides and whose hides were lumpy with maggot-filled abscesses.

We explained that we were looking for birds and mammals and, in particular, for armadillos. We would pay well for any they brought to us and we would give good rewards to anyone who could show us inhabited nests and holes.

As Sandy talked, they looked at us pensively and sucked maté. No one seemed very enthusiastic. I could hardly blame them. It was so hot and humid that lying in a hammock was very much more pleasant than rushing around in the forest. I also realized, with something of a shock, that there were no insects biting us. I interrupted Sandy's exhortations and asked him to inquire if they were ever troubled with mosquitoes, mbaragui or polverines. The villagers slowly shook their heads. I wondered how long my own energy would last if I lived here permanently. Perhaps I, too, if I had no insects goading me and were untroubled by the need to be energetic in order to survive in a competitive society living in a cold climate, would also take to my hammock and wait for the chicken to lay an egg and the bananas to ripen on the tree outside.

The headman explained gravely that our request had come at a rather inconvenient time. For the past few weeks, the men of the village had been discussing whether or not to chop down a tree in the forest nearby which contained some wild honey. They might make up their minds to do so any day now and, obviously, no one could contemplate doing anything else until this particular issue was resolved.

However, he was sure that if anyone happened to *come across* anything, they would collect it if they could and they would let us know. Sandy and I returned to Ihrevu-qua. I could not see much hope of getting any real help from the villagers.

Day after day we roamed through the forest. It was an oppressive, slightly terrifying place. An English wood is gentle and welcoming. Its boundaries are broken by innumerable entrances which invite you to stroll down the sun-dappled corridors that lead to its heart. But the forests that encircled Ihrevu-qua resisted entry with vicious snatching thorns and tangled creepers. When we forced our way inside, we were met with fresh hordes of stinging insects, with ticks and with leeches. If we did not take a compass we had no sure idea of our orientation, for the sun was hidden from view by tier upon tier of leafy screens. In order not to lose our way we marked our route by slashing the tree trunks so that we had a line of white wounds which we could follow to lead us back to safety. All around us were the signs of frantic growth, and of decay and corruption. Most plants must rise to the sunlight to survive; some had outgrown their strength in the effort to do so, had toppled and were rotting on the ground. Creepers and lianas were climbing upwards by clinging to the trunks of established trees and those that had reached their goal were already strangling their helpers. Only where a giant tree had fallen could a shaft of sunlight flood the forest floor, and then there had sprung up a tangled mass of smaller plants which would flourish until such time as new sapling trees overtopped them to steal their light and ultimately kill them. Except in these clear areas we saw few flowers.

There were no big animals in the forest. The largest we could expect to find was the jaguar. It is not rare, but it moves so quietly and its camouflage is so perfect that the traveller seldom sees it unless he hunts it with dogs. Indeed, at first sight the forest seemed deserted, except for the butterflies and the singing, chirping insects which filled the moist air with a continuous chorus of whistles and chirps.

But animals were there, watching us, unseen from their leafy concealments. Once we saw a racoon but it flashed away in front of us and disappeared in a scuffle of rustling leaves. We could only be sure of what it was that we had seen by examining the footprints it had left behind. The ground, indeed, was the ledger in which we could check the identities of the creatures which had been along our path before us and which had silently vanished before we had arrived. The commonest spoor was that of the tegu lizard – a serpentine twisting groove made by its tail with, on either side, the marks of its clawed feet. Sometimes we followed such a trail and saw the lizard itself, gunmetal grey, nearly three feet long and motionless as a statue. But that, too, would be gone in a flash if we approached to within a few yards of it.

Of all the inhabitants of the forest, the birds were the most visible. Trogons, the size of cuckoos, with scarlet chests and bristling moustaches around their beaks, sat in the trees bolt upright by the globular brown termite nests in which they habitually make their own nesting holes. On the ground, the almost flightless tinamou, a little chestnut-brown bird rather like a partridge, stepped hesitantly and unobtrusively through the shade, sometimes calling melodiously with a sustained, liquid whistle. Once we found its nest, filled with a dozen purple eggs as polished and shiny as billiard balls. The urraca jays usually came to find us, for they are so unfailingly inquisitive that if we walked near a party of them they would hop and flutter through the branches towards us and stay close by, shrieking and cackling; they were beautiful creatures with cream undersides, bright blue backs and wings, and curious close-feathered heads which made them look as though they were wearing some odd cap. One bird – the bellbird – we seldom saw, but we knew it was abundant for wherever we went we heard its astonishing metallic call. When we did catch sight of it, we saw no more than a white speck perched on the topmost twig of the highest tree. The bellbirds divided the forest into their own territories and proclaimed their ownership by calling almost continuously for periods of over an hour at a time. Sometimes, too, one would indulge in vocal battle with another half a mile away so that its call seemed to have an echo and the whole forest rang with the cries.

With the arrival of Cayo's passengers, logging work began. The men in pairs went out each day to fell the gigantic hardwood trees, some nearly a hundred feet high. Other men, supervised by Nennito and aided by the two Amerindian boys who lived at Ihrevu-qua, began the laborious job of hauling down the cleaned and weathered logs which had been felled the previous season. They used jinkers – giant wooden wheels, over ten feet in diameter and joined in pairs by heavy wooden axles. The logs were shackled by chains beneath the axles and then pulled out of the forest by a team of specially trained oxen. They were stacked in a clearing on the river bank just below

Bringing out logs with oxteam and jinkers

the huts until such time as there were enough to chain together to form a string of rafts. Then the timber-men would ride them down the river to Asunción, a journey which might take them a month.

After a few days, we sent one of the Amerindian boys to the tolderia to see whether anything had been caught. He came back with exciting news. The chief had captured a toucan, an anteater, three tinamou and best of all, an armadillo – and how much would we pay? I was ashamed of myself for having doubted his enterprise. If the Amerindians were, after all, such energetic hunters it was obviously better that we should leave Ihrevu-qua and camp by the tolderia so that we could be on hand to take charge of the animals as soon as they were caught. The proposition was made even more attractive when I remembered that the Amerindians' valley was almost free from biting insects. Nennito lent us two of his horses on which we loaded our gear. 'If Cayo arrives, send us a message,' we said, 'and we will come back immediately.'

Then we left in high spirits.

We arrived at the tolderia in the evening. The chief, however, was not there. One of the men told us that he was away looking at one of his cassava patches in the forest.

'No, no,' I said with an attempt at mild humour, 'he is looking for more animals for us.'

The villagers laughed rather more heartily than I felt the pleasantry warranted and left us to finish making camp.

The next morning a messenger arrived from the chief.

'The chief has a sore foot,' he announced. 'He cannot come and see you.'

'But the animals,' we said. 'Where are they?'

'I will ask him,' said the messenger and strolled away.

Later that evening the chief appeared. He did not seem to be limping.

'The *señors* wish to pay for the animals,' Sandy said. 'Where is the armadillo?'

'It escaped.'

'And the giant anteater?'

'It died.'

'And the toucan?'

A slight pause.

'It was eaten by a hawk,' said the chief, sepulchrally.

'And the tinamous?'

'Aha,' said the chief. 'I never actually *caught* those but I think I know where I can find them. I only said I'd got them to see how much you would pay.'

Exactly why the chief had claimed to have captured animals he had obviously never possessed was not clear. My own explanation was a rather vague one involving the importance of politeness and the shame of losing face in a primitive society. Charles had a rather more down-to-earth solution.

'I expect,' he said dourly, 'that it was to teach us not to ask silly questions.'

Nevertheless our presence by the tolderia seemed to stir the locals into some activity. It was not enough to stimulate them to begin actually catching animals, but they became quite sympathetically concerned about the success of our trip and often came to our camp to sit about, drink maté and make helpful suggestions as to what we might do and where we might look. One man recalled that he had heard of someone who had recently found some eggs of a bird called *djacu peti*. These, he said, were very rare creatures and the man had taken them home to hatch them underneath one of his domestic hens. From his description it seemed that *djacu peti* was a white-crested guan, a turkey-like bird and one of the most handsome members of its family. We were interested. Where could we find this man? He looked sly. What would we give for the chicks? We haggled for some time and at last we agreed on a scale of barter which was to be finally fixed when we saw how many chicks were for sale, what sort they were and whether they were really healthy. The Amerindian, who clearly had advanced ideas of how much profit

White-crested guans

a middleman was morally justified in making, said that he himself would fetch them.

He left us and returned two days later. The chicks were delightful little balls of fluff, mottled yellow and black. We had no means of knowing whether or not they were white-crested guans, but took his word for it and exchanged them for a knife.

They were already quite tame and they became even more so. Soon they began to follow us around so closely that we were frightened of stepping on them and we had to keep them in an improvised pen for their own safety. They fed with enthusiasm on grain and small pieces of meat, and they grew rapidly. We watched them intently. What would they turn out to be? As time went by, one of them seemed to be slightly different from the others, but it was only some weeks after we got them back to London that we were at last certain of their identity. Three of them were indeed white-crested guans — their wings were black dappled with white, and they had magnificent caps of long white feathers and brilliantly coloured dewlaps, part purple, part scarlet. The fourth bird, however, was much more drably coloured. It was brown with only a small red dewlap. If the Amerindians had purposely sold us this chick from another nest, under the impression they were fobbing us off with an inferior bird, they were mistaken. For this was another species, Sclater's guan, which the London Zoo had seldom had before. For us it was the rarest and most valuable of the quartet.

The sight of a solid shining knife being exchanged for four small guan chicks did not pass unnoticed, and two days later one of the younger men from the tolderia arrived at our camp holding an enormous tegu lizard, fully three feet long, suspended by a noose around its neck. I handled it with the greatest care for the tegu has extremely powerful jaws and I had no doubt that it could easily amputate my finger if I gave it the chance. I gripped it by its neck and its tail. The reptile twisted, there was a faint cracking noise and to my astonishment I found that it had severed its tail close to its back legs and I was holding half the reptile in each hand. The tail was wriggling as energetically as the front half. There was no blood, except for tiny, scarlet pin-points at the end of the long, leaf-like flakes of muscles which projected in a ring around the broken edge. Smaller lizards often shed their tails in this way but to have such a comparative giant do so as I held it in my hands was both unexpected and a little unnerving.

The tegu seemed none the worse for its self-inflicted mutilation, but it had spoilt its beauty. I rewarded the man, but I released the lizard in the forest to let it grow a new tail.

The next day, the same man brought in a second tegu. It was almost as big as the first and I handled it with even greater care. Unfortunately it was injured, for when the man had cornered it in a hole, it had attacked and bitten its captor's bush knife. As a result, its mouth was badly bloodied. I could not believe that it would survive, but I carefully put it in a cage and gave it an egg to eat.

The egg had gone the next morning and the tegu lay somnolently in the corner. Over the next few weeks its mouth slowly healed, and when we eventually handed it over to the London Zoo it was fully recovered and as vicious and spiteful as it had ever been.

Our collection was now quite large. In addition to the guans and the tegu lizard, we had a pair of rare Maximilian's parrots, a young urraca jay and five tiny parrot chicks. But we had still not found the creatures that I was most anxious to see – the armadillos.

Day after day, we searched for their holes. There was no difficulty in finding these as the armadillo is an enthusiastic and energetic hole-digger. It tunnels in search of its food, it digs lots of spare boltholes throughout the forest, no doubt believing that they will come in useful some day. It sometimes abandons its old nesting hole and excavates a new one.

At last we found a burrow that showed every sign of being inhabited. There were fresh footprints near the entrance and scraps of green, unwithered leaves in the rubbish just inside. If there were indeed armadillos in the warren, then the obvious way to catch them was to try to dig them out. I doubted very much if we should be able to capture adults by this method, for they would certainly retreat to the deepest shaft, which might go down as much as fifteen feet, and even if we ourselves could dig as deep as that, I was sure that the armadillos could dig even deeper and much faster than we could. Our real hope lay in the possibility of finding youngsters, for armadillos usually make their nurseries quite close to the surface and avoid the deeper parts of the warren, which are unsuitable for permanent habitation as they are apt to become waterlogged in wet weather.

It was very hard work and extremely hot. The ground was matted together with a tangle of roots. After an hour of exhausting digging we found that the main tunnel was running more or less horizontally about three feet below the surface of the ground. We came across increasing quantities of leaves. I felt sure that

Searching for armadillos

we were getting near to the nesting chamber. On my hands and knees, I cleared away the loosened earth and peered down the tunnel in an attempt to make sure that there was nothing dangerous inside before I put my hand down it; but I could gain very little confidence for I could see nothing whatsoever. The only thing to do was to try. I lay flat on my stomach in the pit which we had excavated and plunged my hand into the hole. I could feel only leaves. Then there was a movement. I made an unseeing grab and caught hold of something that was warm, and wriggled. I was sure that I had an armadillo by its tail, but whatever it was, I could not pull it out. The animal seemed to be bracing its back against the roof of the hole and digging its feet in the floor. I hung on while I managed to squeeze in my other hand. While I was fumbling and struggling, I discovered one fact about the animal I was tackling. It was ticklish. Inadvertently, I touched it under its stomach with my left hand, and as soon as I did so, it doubled up, lost its grip and out it came like a cork from a bottle.

To my joy and relief, I found that I had caught a young nine-banded armadillo. There was no time to examine it in detail, for there might be others inside. Rapidly I put it in a bag and returned to the hole. Within ten minutes, I had caught another three. This was exactly the number I had expected, for the female nine-banded armadillo has the extraordinary characteristic of giving birth to identical quadruplets. We took the four brothers back to our camp in triumph.

An armadillo heading into its burrow

Our first task was to provide them with comfortable cages. Fortunately we had still not used four boxes which had been given to us by a British friend in Asunción and which we had brought with us, dismembered and tied into a neat bundle. Rapidly we reconstructed them and nailed fine wire netting over the tops. With some earth and dried grass inside them, they made perfect cages for the armadillos. They also provided each of the animals with a name, for the boxes had originally been sherry cases and soon we were automatically talking about their occupants as Fino, Amontillado, Oloroso and Sackville. Collectively we called them the Quads.

They were the most attractive creatures. Their shells were pliable but polished and smooth. They had small, inquisitive eyes, and large pink stomachs. For most of the day, they lay sleeping beneath their hay, but in the evening they came to life and rampaged around their boxes, impatient to get their food. And they had enormous appetites.

Nine-banded are the commonest and most widespread of all armadillos. Paraguay is almost their southern limit, but they occur in most of the other South American countries to the north, and during the last fifty years they have extended their territory into the southern part of the United States of America. The Amerindians often came to look at the Quads and would sit on their haunches watching the animals' every movement. Why they should be so interested I could not understand, for all of them must have seen many armadillos before. Indeed the animal forms a much relished item of their diet. Perhaps it was because they had seldom watched a living one for any length of time, for doubtless whenever they captured one they killed and ate it immediately.

They told us many things about them. They said that when an armadillo wanted to cross a river, it simply walked down the bank into the water and then continued walking, submerged, on the riverbed until it reached the other side. It seemed a fantastic story and I did not pay much regard to it. When we got back to England, however, I discovered that this tale is probably quite true. The armour-plating that the armadillo carries on its back makes it very heavy so that it would have no difficulty in remaining on the bottom of the riverbed. Furthermore, it has an astonishing ability to hold its breath for a very long time and to build up an oxygen-debt in its tissues. This is very necessary for it often has to dig very rapidly and continuously, and when it does so its nose is inevitably buried in the ground so that it is almost impossible to draw breath. These two characteristics make it quite possible for an armadillo to walk under water, and an American research worker has been able to encourage them to do so under laboratory conditions. So far, however, there have been no firsthand reports published by a scientist who has observed the armadillos crossing rivers by this method of their own accord and it is also known that they are able to swim in the normal way along the surface if they wish to do so, by filling their lungs with air in order to lighten their heavy bodies.

Now that we had caught the Quads, we began to worry again about our return trip. There had not been any heavy rain during the past few days and it might be that the river was beginning to fall. Cayo might well be on his way down again. It would be disastrous if we missed him, so we gathered all our things together and marched back to Ihrevu-qua.

Nennito and Dolores welcomed us with maté. We sat round the fire passing the gourd from one to another, while we heard the latest news.

The polverines had been very bad during the last few days. The logging was going well and many of the trees had been felled and a lot of logs were lying stacked on the bank. Soon there would be enough to start building the rafts.

'And Cayo?' I asked.

'Gone,' replied Nennito in Spanish, in an offhand way.

'Gone?' We could not believe our ears.

'*Si, si.* The river is getting very low. I asked him to wait while I sent a message to you, but he said that he was in a hurry.'

'But how will we get back?'

'I think perhaps there is another boat upstream somewhere. If there is, I expect they will be coming down some time. I am sure they will take you.'

There was nothing we could do except wait and hope.

Very fortunately, we did not have to wait long. Two days later, a tiny launch came noisily chugging downstream. There were five people on board and obviously no room for us, but the captain agreed to take most of our baggage and the animals. They too were in a hurry. The river was falling fast. If they did not get to Jejui within three days, they said, they might be stranded for weeks until there was more heavy rain and the river rose again. They were not, however, going to Asunción, but only as far as Puerto-i. We reckoned we had more chance of finding an Asunción-bound boat there than in Ihrevú-qua. Within an hour, we had packed up everything, said goodbye to Nennito and Dolores and were in our speedboat, following the launch.

It took us just over three days to get back to the Jejui.

As we approached Puerto-i, I saw another launch coming towards us. I reached for my binoculars. It was the *Cassel*. I could even see the unmistakable, straw-hatted figure of Capitan at the wheel. I would never have imagined that I could be so glad to see him again.

We drew alongside. Gonzales leaned out and waved us on board. As we transferred our gear and the animals, Capitan told us that when he had returned to Asunción the kindly people at the meat factory had been horrified to see him come back without us and had told him to refuel and to return upriver to wait for us. And here he was.

The cabin seemed like paradise.

Charles turned on the radio, lay back on the bunk and started to make a plateful of elegant *canapés* – buttered cream biscuits garnished with neatly curled anchovies.

He took a long drink from a glass of beer at his side. 'Not a bad trip,' he said, meditatively. 'Apart from one or two dodgy days in the middle, not a bad trip at all.'

24

Nests on the Camp

———

The *Cassel* reached Asunción and slid alongside the meat company's jetty early in the morning. Capitan shouted to Gonzales to stop the engine and then clambered ashore, with the broadest smile we had ever seen on his face, to be greeted by his stevedore friends as a returning hero. Gonzales followed him and collected his own attentive audience as he described with lavish gestures the story of the voyage.

After the trials and uncertainties of the past few weeks, Charles and I were equally happy to see once again the garbage-strewn waters and the squalid quays of the Asunción docks. As we walked up to the manager's office to thank him for the use of the launch, my mind was full of the delights that awaited us in the town – a waterproof bedroom, a soft mattress, letters from home, and delicious meals which neither of us had prepared, served on a polished mahogany table and eaten with shining silver cutlery. We should be able to spend at least a week in lazy comfort because, in view of the doubt about the date of our return, we had made no arrangements for another journey and it would inevitably take some time to do so.

The manager greeted us warmly.

'You have got back at just the right moment. You remember that you said that you would like to visit one of our estancias some time? Well, the day after tomorrow, the company's plane is coming to Asunción and it could easily take you down to Ita Caabo on its way back to Buenos Aires, if you would like to go.'

Even though it meant forfeiting our week of luxury there

was no question of refusal, for when Ita Caabo had first been described to us we both realized that a visit there might be one of the most rewarding excursions that we could make. The estancia lay two hundred miles to the south, in Corrientes, the northernmost province of Argentina. For many years it had been managed by a Scotsman, Mr McKie, who believed that the successful ranching of cattle did not automatically necessitate the extermination of all wild animals, and as he was an enthusiastic naturalist he prohibited hunting on the vast area of land under his control. As a result the estancia not only produced great quantities of beef but also became an animal sanctuary. The tradition had been continued by Dick Barton, the present manager, and at Ita Caabo, the wild creatures of the Argentinian plains flourished in greater numbers than almost anywhere else.

Our week of idle living became transformed into two rushed days of feverish activity. We dispatched our exposed film to London and overhauled all the equipment. We built pens and cages as semi-permanent accommodation for our animals in the large garden belonging to the British friends with whom we were staying. To look after the collection in our absence, we engaged, at our hosts' suggestion, their gardener, a delightful Paraguayan lad named Appolonio, and arranged for one of his brothers to come to the house to take over the work in the garden. Appolonio had an abiding passion for animals and the joy and excitement with which he took into his care the guan chicks, the parrots, the Quads and even the surly tegu lizard left us in no doubt that he would tend them all with the utmost devotion.

The company's plane arrived on schedule and proved to be a tiny single-engined aircraft, so small that it was only with considerable difficulty that we managed to cram inside it the basic essentials of our equipment.

Within a few minutes of taking to the air, Asunción and Paraguay had disappeared behind us and we were flying over Argentina. It was a new land, geographically as well as politically. Nothing interrupted the geometric precision of the roads and fences which here and there crossed the grass plains like red and silver lines ruled across a blank green canvas. It seemed almost inconceivable that any wild creature could exist in such a country so devoid of cover and so completely dedicated to the scientific production of beef. For nearly two hours we droned on. Then the pilot yelled to us above the noise of the engine and pointed ahead at a small hollow rectangle of red buildings, enclosed by a narrow belt of trees like a picture in a dark green frame. This was Ita Caabo. The horizon tipped, the buildings loomed larger, and the tiny specks which dotted the plains resolved themselves into cattle. We levelled out and landed.

The manager was awaiting us. He was a tall humorous-faced man, wearing a misshapen trilby hat and leaning on a walking-stick, who might easily have stepped out of a farmhouse in Herefordshire. His first words were as English as his appearance.

'Good afternoon. My name's Barton. Come along in – I'm sure you chaps want a glass of ale.'

The garden through which he led us, however, was far from English. A giant palm tree waved its fronds lazily in the middle of the velvet lawn; jacaranda, bougainvillea and hibiscus blazed in the shrubberies, and working among the flowerbeds, meticulously snipping off the dead heads of some of the plants, stood the romantic figure of an Argentinian cowhand, complete with baggy trousers, a massive leather belt, in which he had stuck a great naked knife, a wide-brimmed hat and a heavy black moustache.

The house itself, single-storied and rambling and roofed with corrugated iron, was hardly beautiful, but though it lacked elegance it was certainly luxurious, being built and furnished on a scale which was almost Edwardian in its opulence. Charles and I were shown into a separate spacious guest suite with its own private bathroom, and we joined Dick Barton in a vast billiard-room for our promised beer.

We told him what animals we hoped to see – rheas, capybara, turtles, armadillos, viscachas, plovers and burrowing owls.

'Bless me,' he said. 'That's easy. We've got lots of 'em. You can have one of our trucks and wander about the place until you find 'em. Anyway, I'll get the men to look out for things and warn them that I shall be fearfully egg-bound if they can't show you what you want.'

Peones at work, Ita Caabo

The land surrounding the house was not entirely flat, as it had appeared to be from the air, but undulating like the gentle sweeping downs of Wiltshire. Nor was it totally devoid of trees, for a few spinneys of Australian casuarinas and eucalyptus had been planted to provide the cattle with shade. Dick spoke of it not as 'the pampas' – that country lay several hundred miles to the south towards Buenos Aires and is as level as a table – but as the 'camp', an anglicized abbreviation of the Spanish word which means, simply, 'countryside'.

The eighty-five thousand acres that belonged to the estancia were divided by wire fences into several vast paddocks, each of which was as large as a small English farm. Their lush grass made excellent grazing for the cattle, but apart from the few copses of shade trees they provided no cover to shelter birds nor any sites for nests. Nonetheless several kinds of birds managed to flourish by employing techniques of nesting specially suited to this open inhospitable country.

The alonzo or ovenbird, a small reddish brown creature about the size of an English thrush, makes no attempt whatever to conceal its nest from hawks or to place it beyond the reach of nuzzling cattle. Instead, it shields its eggs and young from danger by building an almost impregnable nest, a domed construction of sun-baked mud, shaped like the earthen oven in which the local people bake their bread. It is about a foot long and has an entrance which is large enough to admit a man's hand. But the eggs are well protected for, just beyond the entrance, the nest is divided by an internal wall enclosing the nesting chamber which is pierced only by a small hole, just large enough for the bird itself to squeeze through.

An ovenbird with its half-built nest

Having devised such an efficient fortress, the ovenbird has no need to conceal it and builds in the most conspicuous places. If there are no trees available, then it makes its nest on the top of fence posts, telegraph poles, or anything else which will support it above ground level where it might be kicked and cracked by the cattle. We found one which had been cemented on to the top bar of a frequently used gate so that it must have been swung through ninety degrees several times a day.

An ovenbird by its completed nest

The alonzos are bold creatures and seem actively to prefer human company, for they frequently choose to build close to a house. The cowhands, the *peones*, in return for the compliment, are very fond of this small bird which is so companionable and so fearless, and they have given it many pet names. Just as we talk affectionately of Robin Redbreast and Jenny Wren, so they refer to the ovenbird as *Alonzo Garcia* and *João de los Barrios*, which means roughly Mud Puddle Johnny. They say that the bird has an exemplary character: it is cheerful, for it sings constantly; it has high moral principles, for it mates for life; and it is extremely industrious, working from dawn to dusk during the time it is building its nest – except, they say, on Sunday, for it is also exceedingly pious.

In the folds of the downs and on the banks of the streams there grew occasional patches of a thistly weed called caraguata which sends up tall fruiting stems over six feet high from basal rosettes of savagely spined leaves. These thickets were the home of many small and beautiful birds which only occasionally ventured out on to the open camp.

Troops of scissor-tails came there to feed, flying from stem to stem in erratic swoops, or clinging to the top of a particularly tall plant to chirrup their clicking percussive song in the sunshine, opening and closing their long cleft tails as they did so. There also we found the little widow tyrant, pure white except for the tip of its tail and the primary feathers of its wings which are black, and the exquisite churinche, black on its tail, wings and back but elsewhere a miraculous vivid scarlet. The peones call it, in Spanish, the Fireman, Bull's Blood, or, perhaps most appropriate of all, *Brazita del Fuego*, the Little Coal of Fire. Whenever we saw the churinche, we had to stop to gaze on its beauty and mourn the fact that the film we were making was in black and white.

The most elegant of all the inhabitants of the camp were the rheas. Dick considered that we were fussily pedantic to call them anything other than ostriches and, indeed, the two birds are very similar. True ostriches, however, live only in Africa and the rheas, which are restricted to South America, differ in several details. They are a little smaller, their feathers are not black and white but a warm ashy grey, and they have three toes on each foot, whereas the ostrich has only two.

317

We saw the rheas often, pacing slowly over the grasslands with the dainty deliberation of mannequins. The estancia's ban on hunting had made them comparatively fearless and they allowed us to drive within a few yards of them. They warned us when we had reached their safety limit by ceasing to graze and lifting their heads to stare suspiciously at us in the same way as deer will do. Their long necks inevitably gave them a supercilious look, but their large eyes were liquid and gentle.

Being flightless birds, their fluffy billowing wings served no function, except perhaps to keep them warm, for their bodies are only scantily clothed in short cream-coloured feathers and when they ruffled their wings, wrapping them around their naked-looking bodies, they did so with the air of rather chilly fan dancers.

Each group was made up of one male bird and a number of females of varying sizes and ages. The male was usually the largest and was distinguishable from his wives by the black stripe which ran down the back of his neck and in a narrow yoke around his shoulders. The females also possessed this stripe, but it was brown and not so distinctly marked.

If we ignored their warning gaze and drove too close to them, then the whole party would stampede away, running leggily and at great speed, their powerful feet beating a muffled tattoo on the ground. Dick told us that they could out-distance all but the speediest of horses and were so adept at swerving and jinking that they were extremely difficult to catch.

We found one of their nests in a marshy patch of reeds. It was a shallow depression, nearly three feet across, lined with dry leaves, and it contained an astonishing number of gigantic white eggs, each nearly six inches long and over a pint and a half in capacity. There were thirty of them, lying untidily in the nest. As I looked at them, I made a rough mental calculation. In terms of yolk and albumen, I had found in this one nest the equivalent of nearly five hundred chickens' eggs. However this was not an exceptionally large nest. The previous season a peon had found one which contained fifty-three eggs, and W. H. Hudson wrote of a really gargantuan clutch of one hundred and twenty.

Needless to say all the eggs in a nest are not laid by a single female. All the members of the male rhea's harem make their contributions and, when I looked closely at the nest we had discovered, I could see that the eggs varied slightly in size, the smaller ones having been laid by the younger females.

A rhea's nest

A number of questions formed in my mind. I knew that the male selected the nesting site and incubated the eggs after they had been laid, but how did all his wives know where he had built the nest and how were the laying operations organized so that all the females did not wish to lay at the same moment, or alternatively that no eggs were added to the nest for days on end? Unfortunately, we could not find the answers to these questions by keeping watch on this particular nest, for the eggs were cold. It was deserted.

Three days later, however, as we walked into a patch of cara-guata growing on the banks of a stream, to get a closer view of a churinche, a rhea sprang up in front of us and thudded away, dodging and weaving through the tall stems. We found his nest a few yards in front of us. It contained only two eggs. If we kept continuous watch on it we might perhaps be lucky enough to see exactly what did happen when the rheas came to lay.

From previous experience we decided to use the car as a hide. The best vantage point was some thirty yards away on a rising slope which would enable us to look down slightly on the nest. The caraguata, however, was so thick that the nest was invisible at a distance of more than a few feet. Carefully, we cut down a few of the taller stems to form the beginnings of a narrow corridor down which we should be able to look. I was anxious not to trim too much at one time, so that the male rhea would become accustomed bit by bit to what might amount to a considerable change in the surroundings of his nest.

Every morning, for the next few days, we returned to the site, each time parking the car in exactly the same position. Every morning, as soon as the rhea left his nest, we enlarged and perfected the narrow avenue leading down from the car to the eggs. We knew that our activities were not disturbing the birds, for every morning we saw an extra egg, newly laid and bright yellow, contrasting strongly with the remainder of the clutch which had faded to an ivory colour. At last our corridor was clear and on the fifth morning, we began our watch.

By now, we felt we knew the old male rhea quite well. We christened him Blackneck. He was sitting on the nest, but in spite of all the gardening we had done it was still quite difficult to distinguish him, for his grey feathers matched and blended with the grass and caraguata, and he had folded his long neck so that his head lay on his shoulders. Only his bright eyes gave a clue to his presence and even then, had I not known exactly where to look, I do not suppose that I should have noticed him. We settled down for a lengthy stay.

Two hours later, Blackneck had still done nothing. He had barely moved. The sun had risen and already it was getting hot. Some cows that had been grazing on the open camp when we arrived had retired to the shade of a plantation of eucalyptus trees behind us. Beyond the nest, a heron which had been fishing in the stream flapped noisily away, its breakfast finished. Blackneck sat motionless. Every few minutes, I raised my binoculars in the hope that I might see him doing something interesting. The only movement he made was to blink.

We had now been watching from the car for two hours. Surely the bird could not have started incubation yet. He could not have more than six eggs beneath him – nothing like a full clutch. Over the brow of the hill on our right a group of six rheas appeared, grazing idly. They were all females – his harem. They worked their way slowly towards us, then back again to disappear over the skyline.

Blackneck got to his feet. He stood for a moment and then slowly walked away in the direction of his wives.

We then began a long wait. Blackneck had left at nine o'clock and we saw nothing of him or his harem for the next three hours. Then, at a quarter past twelve, he strolled over the brow of the hill accompanied by a young female. The two of them walked together towards the nest. I would have liked to have imagined that Blackneck was leading or shepherding his partner towards the nest, but it was impossible to tell if that were so. However, as there were more wives in his harem than there were eggs in the nest, it was quite likely that she herself had never previously visited the nest, in which case Blackneck must have been showing her where it was.

Whether or not she had seen the nest before, when she finally reached it she did not seem to think very highly of it. For several minutes she examined it closely. Then she bent her head and picked up a small feather from among the eggs and disdainfully flicked it over her shoulder. Blackneck stood by her, watching. She made one or two other alterations to the nest, but even when she had spent some time tidying it up, it still did not seem to meet with her approval for she stalked away to the left, through the tall caraguata stems. Blackneck followed her.

They walked on for a hundred yards until, suddenly, she sat down, almost disappearing in the tall grass. Blackneck, who had been leading, turned to face her – and us – and began to sway his head from side to side. The actions of most mating displays seem designed to show off the markings and adornments which distinguish sex, and certainly Blackneck's dance had the effect of flaunting his glossy black neck stripe and yoke in front of his mate. Almost immediately Blackneck took a step towards her. Their weaving necks came closer and closer until they met and, snake-like, entwined. For a few seconds they swayed together ecstatically. Then the female sank down to the ground again. Blackneck broke away and in a flurry of grey plumes he mounted her, his head held low. They remained so for several minutes and all we could see of them was a grey hump of feathers. Then they separated and Blackneck walked away up the hill, nibbling in a desultory way at some caraguata fruits. The female rose to her feet and joined him, both of them ruffling their wings to set them back neatly into position. They came back together to the nest. Once again the female stooped and examined it, but she did not sit and the pair walked away to the right in the direction of the rest of the harem.

Once more the nest was vacant and there were no rheas in sight. We sat on in silence, doggedly determined to see an egg being laid. It seemed clear that we had seen the first part of the process. We had witnessed the male bird showing the nest to one of his wives and then coupling with her. If this had been the initial mating with this particular hen, then she would not be ready to lay a fertile egg for some days. It might be, however, that the mating we had watched was the continuation of the display and designed to serve as a stimulus to lay. We had no means of knowing.

Blackneck with one of his females

For three hours, the nest was deserted. Then at four o'clock a female appeared again from the caraguata thicket on the right, and behind her came Blackneck. They walked straight up to the nest. We could not tell whether this female was the same bird that we had seen earlier in the morning or another one. She examined the nest, removed pieces of dried leaves from it and then, very slowly, head and neck erect, she sank down.

It had never occurred to me to wonder what a male bird did while its mate was laying an egg. Most males, I imagine, are not present at the time and are quite oblivious of its happening. Not so Blackneck. He strode up and down behind the nest as the female sat, looking as agitated as any father outside the maternity ward of a hospital. The female did not seem comfortable. She ruffled her wings once or twice. Then she lowered her head to the ground. Several minutes later she rose and rejoined him, and together they walked away.

When they had gone, I quietly got out of the car and walked down to the nest. There, outside the nest's rim, was a seventh egg, still wet and a bright yellow. The hen's body was so large that the egg had been laid well away from the other. No doubt Blackneck would be back as evening came to put it into place with the others and guard the clutch for the night.

We started up the car and jubilantly drove back to the house. We had found the answer to at least one of my questions. It was the male rhea who showed the females the position of the nest and he who organized the egg-laying.

But there was one story we had not been able to verify. Sandy Wood had told us that when the male bird has a complete clutch and begins to incubate, he pushes one of the eggs outside the nest. This he said was called *el diezmo* – the tithe. This stays by the nest until the main clutch hatches. Then the male bird breaks it open with a kick so that the yolk spills on the ground. Within a few days, this patch of earth is wriggling with maggots, which provide the young chicks with a perfect meal just when they are in most need of it. I wished we could have stayed long enough at Ita Caabo to have seen Blackneck do so.

25

Beasts in the Bathroom

For the animal collector, there is no more useful room than the bathroom. I had first discovered this truth in West Africa, when we stayed in a rest-house the bathroom of which was so primitive that we had little compunction in foregoing its largely hypothetical amenities and using it as an annexe of our embryonic zoo. Its only claim to its title was a monstrous, somewhat chipped enamel bath which stood majestically in the middle of the otherwise bare floor of red earth. It still sported a plug, shackled to the brass overflow by a heavy chain, and its taps were bravely labelled 'Hot' and 'Cold', but if the water had ever flowed through their tarnished Victorian nozzles, it must have done so in some earlier, more distinguished situation, for here they were unconnected to any pipes and the only running water within miles flowed through a nearby river.

But though that bathroom had little to commend it as a place in which to wash, it provided excellent accommodation for animals. A large fluffy owl chick relished the gloom, so similar to the dim light of its nesting hole, and perched happily on a stick thrust through the rush walls across one corner. Six corpulent toads inhabited the dank clammy recesses beneath the deep end of the bath, and later a young crocodile, a yard long, lounged in the bath itself.

To be truthful, the bath was not the ideal home for the crocodile, because, although he was unable to scale its smooth sides during the day, at night he seemed to draw on extra sources of energy and each morning we found him wandering loose on the floor. We took it in turns, as one of the regular before-breakfast chores, to drop a wet flannel over his eyes, pick him up by the back of his neck while he was still blindfold, and put him back, grunting with indignation, into his enamel pond.

Since that time, we had kept hummingbirds and chameleons, pythons, electric eels and otters in bathrooms as far apart as Surinam, Java and New Guinea, and when Dick Barton showed us the elegantly appointed private one at Ita Caabo, I noted appreciatively that it was by far the most suitable that we had ever had at our disposal. Its floor was tiled, its walls of concrete, the door stout and close-fitting and it was furnished not only with a bath possessing fully functional taps, but a lavatory and hand basin as well. The possibilities were immense.

When we first flew in the company's plane, I had decided that there would be no room in it for any animals on the return trip to Asunción, but as the days passed, and the precise memory of the plane's size began to fade, I managed to convince myself that there must be room for just one or two small creatures. It seemed a criminal waste not to take some advantage of the bathroom's potentialities.

I found the first lodger for the bathroom one day when I was out riding on the camp shortly after a heavy rainstorm. The paddocks were waterlogged and in the hollows, wide shallow pools had formed. As I rode past one of them, I noticed a small frog-like face peering above the surface of the water, gravely inspecting me. As I dismounted, the face disappeared in a muddy swirl. I tied my horse to the fence and sat down to wait. Soon the face appeared again from the farther edge of the pool. I walked round towards it and was soon close enough to see that whatever else this inquisitive little creature might be, it was not a frog. Again it vanished and swam away beneath the surface, stirring up a cloudy line as it went. The trail stopped as the animal settled. I put my hand into the water and brought up a small turtle.

He had a beautifully marked underside, patterned in black and white, and a neck so long that he was unable to retract it straight inwards like a tortoise, but had had to fold it sideways. He was a side-necked turtle – not a rare creature, but an engaging one, and I was quite sure that we could find room in the aeroplane for one so small and attractive, even if he had to travel in my pocket. The bath, half-filled, with a few boulders in the deep end on which he could climb when he was bored with swimming, made him an excellent home.

Two days later, in one of the streams, we found him a mate. As the pair of them lay motionless on the bottom of the bath, each displayed two brilliant black and white fleshy tabs which hung down from beneath their chins like lawyers' bands. It may be that odd appendages, which their owner can move about if it wishes to do so, serve as lures to attract small fish fatally close to the turtle's mouth as it lies unobtrusive and stone-like on the bottom of the pond. But our turtles had no need to use them, for each evening we begged some raw meat from the kitchen and offered it to them with a pair of forceps. They fed eagerly, shooting their necks forward to engulf the meat in their mouths. As soon as they had finished their meal, we took them out of the water and let them wander around on the tiled floor while we used the bath for its more conventional purpose.

I was particularly curious to discover what sort of armadillos lived in this part of Argentina, for it might be that there was one which was not to be found in Paraguay. Dick told us that two different kinds were commonly found on the camp; one was the nine-banded species which we had found on the Curuguati, but the other, which he called the *mulita* or 'little mule', sounded unfamiliar. Dick promised to ask the peones to bring one in if they saw one, and the very next day the foreman arrived at the house with a mulita wriggling inside a bag.

To our delight it proved to be a species that, as far as we knew, did not occur in Paraguay. Although it resembled the nine-banded in general shape, it had only seven bands of separate articulating plates across the middle of its back, and its shell was not polished and shiny, but rough, black and warty. We must surely find space in the plane for him. The Quads had already taught us that armadillos are powerful and persevering burrowers and will demolish all but the most substantial cage. There seemed little point in building one, when the bathroom, tiled, spacious and secure, was still relatively under-populated. We gathered a pile of dry hay, put it in a corner by the lavatory, together with a dish of minced meat and milk, and released the mulita in his new home. He dived straight into the hay, and scuffled invisibly to and fro so that the pile heaved and tossed like a stormy sea. Then tiring of that, he poked his head out, caught a whiff of the meat, trotted over to the dish and began to feed, chewing the meat and puffing so stertorously that he blew bubbles of milk from his nostrils. We watched him finish his dinner. Then we went to bed, happy in the thought that we had a second species of armadillo to join the Quads.

When I went into the bathroom the next morning to shave, the mulita was nowhere to be seen. I assumed that he must be slumbering beneath the hay, but when I looked he was not there. It was difficult to imagine where he might be hiding in the bleak, hygienic bathroom. I looked beneath the bath, behind the lavatory and at the base of the towel rail and the hand basin. That seemed to exhaust all feasible hiding places, but still I could not find him. It was impossible to believe that he had escaped, for there was no way out. The only explanation was that one of the servants had opened the door and inadvertently released him. Dick was most upset at the news and questioned all the servants but none of them had been into the bathroom that morning. We searched again after breakfast. The mulita had certainly disappeared but how and by what route we could not imagine.

Two days later we were brought a second mulita. This one was a female. Once again we put her in the bathroom, and at hourly intervals throughout the evening I looked in to see how she was faring. She seemed very comfortable and fed with as much gusto as her predecessor. But when I went to visit her at midnight, she too had gone. She *must* be in the bathroom somewhere. I called

Charles and Dick and together we began a detailed search. Perhaps, in some mysterious way, she had dived into the lavatory. We lifted the manhole cover in the courtyard outside, but there was no sign of her. We crawled around the bathroom floor looking for unseen gratings or crannies but found none. Then at last, sticking out of the small space between the wall and the base of the lavatory, we discovered a black warty tail. She had gone to earth inside the hollow porcelain pedestal. Getting her out was extremely difficult for she had wedged herself very tightly and we only managed to do so by employing the stomach-tickling technique which we had learned on the Curuguati. When finally she had been extracted Charles peered into the porcelain cavern, marvelling that she had managed to squeeze herself into such a confined space.

He sat back and grinned.

'Have a look,' he said, and there at the bottom of a tunnel almost concealed in the loose soil of the foundations, I saw a black hump. It was the first mulita. Only an armadillo could have discovered this one chink in the bathroom's defences but I was sure that, with a little rearrangement, the room could still successfully house even such expert escapologists as the mulitas. I half-filled the hand basin and transferred the turtles to it. Then I emptied the bath, lined it with hay and put in the two mulitas. They scampered around among the hay, skidding wildly over the smooth enamel. They pushed their noses down the plughole, made one or two tentative scratches at the brass rim, decided that it was not suitable for excavation and then settled down beneath the hay and went to sleep.

We turned out the light and left them.

'You know,' said Dick, 'I'm almost sorry we discovered where they were. I'm sure they would have provided many entertaining and instructive hours for our future guests. After all, there cannot be many bathrooms with resident armadillos in the loo.'

Half a mile from the house, a deep stream wandered across the camp, gliding between banks high with reedy plants and lined with overhanging willow trees. Here and there it rippled between constricting sand banks or splashed white over a natural weir of boulders, but for most of its course it slid gently from one placid sun-dappled pool to another. Herons and egrets came here to fish, standing knee-deep in the shallows; dragonflies flashed their iridescent wings over its surface as they hawked for mosquitoes and midges; and in its more secluded reaches, families of teal floated in pretty squadrons. All these we had seen for ourselves, but Dick told us of one particular stretch where, he said, we should find capybara.

This was exciting news. Charles and I had long tried to film these odd creatures in the wild state, as opposed to the tame ones we had filmed and collected in Guyana.

They are not rare creatures but they are very shy and wary for the capybara is keenly hunted both for its flesh, which has a flavour reminiscent of veal, and for its hide, which makes unusually soft, pliable leather, much prized for aprons and saddle cloths.

'You won't have any difficulty here,' Dick told us confidently. 'There are hundreds and, because no one is allowed to hunt them, they are as bold as brass. Anyone could get a snap of them with a box Brownie, let alone with all the complicated para-phernalia that you've got.'

We treated this remark with a certain reserve. People had said such things to us before and they usually presaged the immediate disappearance of all animal life in the neighbourhood and, as a result, unsympathetic gibes about our prowess as hawk-eyed observers; but the next day, armed with our most powerful telescopic lens and prepared for the worst, we followed Dick's directions and drove down to the stream. We came upon it suddenly, as we rounded a plantation of eucalyptus. Charles cautiously stopped the car and I scanned with my binoculars the line of trees which marked its banks. I could hardly believe my eyes. Even Dick's description, taken at its face value, barely did justice to what I saw.

Over a hundred capybara lay on the grass by the water's edge, as crowded as Bank Holiday bathers on Blackpool beach. Mothers squatted on their haunches, indulgently watching their babies frolicking around them. Old gentlemen snoozed full length, in isolation, their heads sunk on their outstretched forelegs. Young bucks strolled idly among the family groups, sometimes disturbing one of their dozing elders and being forced to make hasty lumbering canters to safety before they became involved in a quarrel. But the air was heavy with heat and most of the herd were in no mood for strenuous activity.

We drove slowly closer. One or two of the old males heaved themselves on to their haunches and gravely inspected us. Then they turned away and resumed their sleep. Their heads in profile were almost rectangular and their shoulders were shaggy with a long reddish mane. On their muzzles, halfway between their nostrils and their eyes, projected a conspicuous weal-like gland, which the females did not possess. They had an air of nobility and a supercilious expression which reminded me not of their rodent relatives such as rats or mice but of lions.

A mother slowly plodded down to the river, her six youngsters following at her heels in Indian file, and waded into the cooling water. We were now close enough to see that there were almost as many swimmers as there were sunbathers. They floated idly or swam nonchalantly back and forth, seemingly with no purpose other than enjoyment. An old female stood belly-deep in the water, meditatively champing lily leaves. Only one among the whole herd, a young male, was swimming fast. We watched him cross the width of the river, a bow wave fanning from the back of his neck. Unexpectedly, he submerged. We followed the line of ripples that marked his course until suddenly he bobbed up, gasping and blowing, beside a sleek female who had been floating demurely close to the opposite bank. She immediately swam away, and the two of them, with only their brown heads showing, forged down the river, like model boats steaming in line. She attempted to avoid him by diving, but he did likewise and when she surfaced again he was still beside her. The flirtation continued up and down the river for ten minutes or more, the male pursuing her with ardour and skill. At last she relented, and they coupled in the shallows beneath an overhanging willow.

We filmed the herd for two hours that morning, and nearly every day that followed we went down to the stream to watch them, for they were a unique sight. Nowhere else in the world can there be so many capybara so close to civilization.

Capybara by the river

Paradoxically, the creature which had once been the commonest in Argentina, a rabbit-like animal called the viscacha, was now, at Ita Caabo, the rarest.

Seventy years ago, Hudson wrote that there were parts of the pampas where a man could ride for five hundred miles and not advance more than half a mile without seeing one of their warrens, and that in some places a horseman could see at least a hundred at one time. This vast population of viscachas had arisen largely because of the actions of the estancieros themselves, for they had hunted and killed most of the jaguar and foxes which are the viscacha's natural enemies, so that it was able to proliferate unmolested. Soon, however, the ranchers realized that these swarms of creatures were consuming so much grass that they were ruining the pasturage, and a vigorous war was begun against them. Streams were diverted to flood the warrens, and the animals, driven above ground by the waters, were clubbed and slaughtered. The *viscacheras*, as the warrens are called, were partially dug out and the tunnels blocked with stones and earth so that the animals, trapped below ground, starved and died. When the hunters used this method, however, they had to guard the ravaged warren overnight, for viscachas from nearby colonies in some mysterious way became aware of their neighbours' plight and, unless prevented, would come to the assistance of their entombed companions and clear the tunnels. Today, there are few viscachas left. Although Dick could have easily ordered their total extermination at Ita Caabo, he preserved one colony in a distant corner of the estancia and, late one afternoon, he took us in a truck to see it.

After half an hour's driving, we left the rutted earth track and bumped over the tussocky turf through tall thistles, to park twenty yards away from a low mound of bare earth capped with untidy piles of stones, dry wood and roots. Around its base we saw a dozen great holes.

The cairns of stones were not part of a naturally occurring outcrop of rock, but had been built by the viscachas themselves, for these creatures are possessed by a collecting mania. Not only do they drag on to the top of their mounds all the stones and roots which they excavate from their burrows, but they also gather up any interesting and movable object that they find on the camp. If a peon loses something when he is out riding, he is most likely to find it in these untidy but cherished museums.

The animals themselves were still below the ground slumbering in their maze of tunnels, for they only come up in the evenings to graze in the safety of darkness.

It was pleasantly cool. A gentle breeze fanned our faces and rustled among the caraguata. A group of four rheas appeared on the skyline and stalked slowly towards us. They sat down on a patch of bare earth and, ruffling their downy wings, lowered their heads to indulge in a dust bath. The cries of the spur-winged plovers dwindled and ceased, and the birds settled in pairs by their nests. The crimson sun, looking gigantic, sank slowly to meet the straight line of the horizon.

Although the builders of the warren had not yet appeared, the mound was by no means deserted. A pair of small burrowing owls, with striped waistcoats and bright yellow eyes, stood bolt upright, like sentinels, on the top of the stones. These birds are quite capable of digging their own nest holes, but often they make use of one of the outlying burrows of the viscachera, and take advantage of the stone cairns by using them as observation posts from which they can survey the surrounding country and find the rodents and insects on which they prey.

These two seemed to own a hole on the far side and were most concerned by our presence, bobbing up and down, swivelling their heads and blinking angrily. Every now and then they lost courage and scuttled down their holes, only to reappear a few minutes later to glare at us once more.

They were not the only lodgers in the warren. Several small miner birds pattered around on the surrounding close-clipped grass. They nest in long narrow tunnels, but as there are few suitable sites elsewhere on the camp they usually excavate in the sides of the viscacha's hole, just beyond the entrance. Like the ovenbird, to which they are closely related, they build themselves a new nest each year, but their old tunnels had not been wasted for they had been taken over by the swallows which were gliding

A burrowing owl by its nest hole

and swooping around the warren. The viscachera was in fact the focal point for most of the wild life in the neighbourhood. While the lodgers disported themselves in the soft evening light, we waited patiently for the landlord himself to appear.

We did not see him arrive, but suddenly noticed that he had materialized and was squatting, like some grey boulder, by one of the entrances.

He looked like a rather large portly grey rabbit, except that he had short ears and a broad black horizontal stripe across his nose, which made it seem as though he had been trying to stick his head sideways through some newly painted railings. He scratched himself behind his ear with his hindleg, and grunted, jerking his body and baring his teeth as he did so. Then he hopped clumsily on to the top of his mound and settled down to survey the world and see how it had changed since he was last up there. Having satisfied himself that all was well, he began a more careful toilet, sitting up to scratch his cream-coloured stomach with both his front paws.

Charles climbed cautiously out of the car and, carrying the camera and tripod, moved a step at a time closer and closer to him. The viscacha transferred his attention from his stomach to his long whiskers, combing them carefully. Charles walked faster, for now the sun was sinking rapidly and he was anxious to get really close before the light faded so much that photography became impossible. Swiftly though he moved the viscacha remained unperturbed and eventually Charles set down the camera within four feet of him. The burrowing owls, aghast, had retreated to gaze indignantly at us from a tussock many yards away. The miner birds flew twittering nervously around our heads. But the viscacha, impervious and unconcerned, remained on his ancestral throne of stones, like royalty sitting for his portrait.

———

Our stay at Ita Caabo was only a short one. Two weeks after we had arrived, the company's plane returned to take us back to Asunción. It had been a comfortable and fascinating interlude and we were sorry to go. We took back with us the mulitas, the turtles, a little tame fox given us by one of the peones, and unforgettable memories and film of ovenbirds and burrowing owls, plovers and rheas, viscachas and, perhaps most memorable of all, the giant herd of capybara.

26

Chasing a Giant

___

From the cobbled hilly streets of Asunción, you can look over the docks, thick with shipping, across the brown expanse of the Paraguay River and into a flat desolate wilderness. It begins on the opposite bank of the river and stretches westwards beyond the horizon and over the Bolivian border, five hundred miles, to the foothills of the Andes. This is the Gran Chaco. For part of the year it is a parched desert of dusty plains and cactus scrub, but in summer it turns into a gigantic mosquito-ridden swamp flooded by the heavy rains and the streams which, swollen by melting snow, pour down on to it from the flanks of the Andes. We had decided that we should spend the remainder of our time in Paraguay in this extraordinary country. Everyone in Asunción had something to tell us about the Chaco. Most people described hideous hardships, others gave us lists of essential but unlikely-sounding equipment, and some produced fairly convincing reasons why we should not go there at all.

On one thing, all were agreed: it was very hot. Accordingly, we began our preparations by looking for two straw hats. We went first to a small shop down by the docks, the window of which, opening on to a shady colonnade, was filled with a wide variety of cheap clothing.

'*Sombreros?*' we asked. Our Spanish, fortunately, was not to be taxed further, for the proprietor, a young but very stout unshaven man, with a jungle of black curly hair, a loosely knotted tie and very few teeth, had once been to the United States and as a result spoke a picturesque brand of Brooklynese. He provided us with the hats which were inexpensive and exactly what we wanted. Unwisely, we told him why we were buying them.

'The Chaco, she's a very terrible bad place,' he said with relish. 'Ho gracious, the mosquitoes and the *bichos* they *muy muy bravo*. They so plenty, you make the snatch in the air and you got yourself one hextralarge-size, foist-class steak. *Amigos*, they gonna *devour* you.'

He paused, entranced by the vision. Then he beamed.

'I got hextra-fine-quality mosquito net.' We bought two.

He leaned conspiratorially over his counter.

'She gets plenty terrible cold,' he said. 'In the night, ho gracious, you gonna freeze. But you no gotta worry one bit. I got fines' *ponchos* in Asunción.'

He produced two cheap blankets, slit in the middle so that you could slip them over your head and wear them as cloaks. We bought them.

'You plenty good on the horse's back? You go like Gary Cooper?'

We had to admit that we did not.

'She don' matter, you gonna learn,' he said hastily, 'an' you gonna need *bombachos*.' He produced two pairs of pleated baggy pantaloons. This seemed too much.

'Not necessary, *muchissima gracias*,' we protested. 'We're going to wear trousers *inglesi*.'

He screwed his face up in a frightening simulation of agony.

'She's not possible, *amigos*. You gonna hurt yourselves real terrible. You *mus'* have bombachos.'

We capitulated. In doing so, we laid ourselves wide open for his next attack.

'Now you got yourselves plenty beautiful, very lovely, high class bombachos,' he said reflectively, as if to congratulate us on our skill in selecting them with such perspicacity, 'but the cactus an' the bush in the Chaco she very spiny.' He clawed the air to make his meaning clear. 'She gonna tear your beautiful bombachos in plenty hundred pieces.'

We waited for the sequel.

'DON' WORRY,' he shouted, and with a flourish, like a magician pulling a rabbit from a hat, he produced two pairs of leather leggings from beneath the counter. *'Piernera.'*

We owned ourselves beaten. We bought them. There now seemed no portion of our anatomy that he had not clothed; but he had not finished. He peered over the counter and dispassionately looked us up and down.

'You no got belly,' he concluded sadly, 'but,' he added firmly, 'I t'ink you gonna need *faja*,' and he reached down from a shelf behind him two rolls of thickly woven material, about six inches wide. 'Look, I show you.' He wrapped one of them three times around his enormous corporation and then indulged in a pantomime, jumping up and down as though he were on the back of a horse.

'You see,' he said triumphantly, 'the guts. They don' bang about.'

Heavily laden and totally defeated, we staggered from the shop.

'I don't know how much of this lot is going to be any use to us in the Chaco,' said Charles, 'but we're certainly going to be absolute knockouts at the next fancy dress ball.'

These curious items of clothing were not the only equipment which we were persuaded was essential if we were to survive our stay in what some people rather melodramatically referred to as *'L'Inferno Verde'* – Green Hell. We acquired special half-length boots without which, apparently, it was impossible to ride a Chaco horse; two dozen label-less bottles filled with evil-smelling yellow fluid which we were assured contained extra strong, army surplus, insect repellent; several lengths of very thick elastic which Charles had discovered in the market and which he had been unable to resist ('Jolly useful, old boy, for traps and things like that'); great quantities of anti-snake-bite serum (together with an appropriately gargantuan hypodermic syringe), which had been pressed on us by one of our kinder but more pessimistic Paraguayan friends; and a wooden crate of tinned foodstuffs that felt, when we lifted it, as though it were filled with lead weights.

Our preparations were now almost complete. We had rediscovered Sandy Wood in the tourist agency for which he worked and had once more enlisted his aid as interpreter, and we had managed to reserve three passages on a plane that was flying out to a remote estancia in the middle of the Chaco.

We now had three days to spare before we left Asunción and we decided to fill them seeking for one of Paraguay's unique riches – its music. When the first Spanish settlers and Jesuit missionaries came to the country three hundred and fifty years before, they found that the Guarani Indians possessed only a primitive form of music – simple and monotonous, slow in tempo and minor in key. The missionaries introduced their converts to European instruments and the Guarani quickly and enthusiastically learned to play them. Their latent musicality soon flowered into a widespread passion. As they absorbed each new European style – the polka, the galop, the waltz – so they transformed it into something fresh and individual, at once rhythmic yet languorous. Furthermore they began to make instruments for themselves. The guitar they adopted unchanged, but the harp they modified into virtually a new instrument. Their version is made entirely of wood; it is small and portable; and unlike the European concert harp it does not have any pedals so that the performer cannot play semitones. But this seems to be no handicap and the Paraguayan harpist exploits all his instrument's potentialities, not only playing melodies with impressive dexterity, but richly decorating them, sweeping his fingers across the strings to produce thrilling glissandos, or plucking the bass strings to add a heady rhythmic beat. I had already heard something of this ravishing music from records made by Paraguayan groups visiting Europe. Now I wanted to hear it played in its proper setting.

One of the most skilled of Paraguay's instrument makers lived and worked in the little village of Luque, a few miles from Asunción, and we went out to see him. His small house was surrounded by the fragrant orange groves that are so typical of this fertile and beautiful part of Paraguay. He himself was seated at his work bench, polishing a part of a harp with the unhurried loving movements which bespeak the true craftsman. Two tame parrots swung in the rafters of the stable behind him and a pet hawk perched on a wooden stand in the garden. We sat down beneath the orange trees, and his wife brought us some cold maté. While we passed it from one to another, he played to us on the guitar he had just completed. Two lads from a nearby farm joined us, and for an hour they played guitars and sang in the bittersweet half strident voices which are so typically Paraguayan. Their music was gentle and full of engaging cross-rhythms and syncopations; it had none

The guitar-maker

of the harsh almost savage rhythm that characterizes the music of neighbouring Brazil, for that is the contribution of the African element in the population and very few Africans have ever lived in Paraguay. Eventually the guitar was passed to me and the old man asked me to play 'una cancion inglesi'. I did the best I could.

The guitar he had handed to me was most beautifully made and had a rich mellow tone. I admired it so much that I asked, in as tactful a way as I could, whether it was possible to buy it.

'No, no,' the old man replied with such vehemence that I feared for a moment that I had offended him. 'I could not allow you to have that one. It is not good enough. I will make a guitar specially for you that will sing like a bird.'

When, a month later, we returned to Asunción after our journey in the Chaco, I was to find the guitar awaiting me. It had been made from the handsome woods of the Paraguayan forests and at the top of its fingerboard the old craftsman had inlaid my initials in ivory.

The next day we met Sandy in a bar in the centre of the town, where he was obviously preparing himself to withstand weeks of total drought in the Chaco. He bought us a beer.

'By the way,' he said, 'a chap came into the agency yesterday asking if it was true that there were some boys in town who were interested in armadillos. He said he had got a *tatu carreta*.'

I nearly choked over my drink. *Tatu carreta* – 'cart-sized tatu' – is the local name for the giant armadillo. It is a magnificent creature, almost five feet in length and so rare that it had never been brought alive to England. Very few people have ever seen a living one and only in moments of wildest optimism had I dared hope that we would be able to find it.

'Where is this man? What is he feeding it on? Is it in good health? What does he want for it?' Excitedly, we bombarded

Sandy with questions. He took a long and contemplative drink of beer.

'Well, I don't rightly know where he is now. If you are as interested as all that, we'll go and find out. I didn't see him myself.'

We rushed over to the agency and found the clerk who had spoken to the man.

'He just wandered in,' the clerk told us, amazed at our excitement, 'and asked how much the *inglesi* would pay for a tatu carreta because he had a captive one for sale. I didn't know whether you would be interested or not, so he said that he would come back some other time. His name, I think, is Aquino.'

One of the loungers, who habitually spent his days sitting on the agency steps, joined in the conversation.

'I think he sometimes works for a timber firm down by the docks.'

In a fever of excitement, we hailed a taxi and drove off to try and trace him. At the offices of the timber firm, we discovered that Aquino had arrived three days ago on board a cargo boat loaded with logs. He had come from the riverside town of Concepción, a hundred miles away to the north, but he had not brought a giant armadillo with him. It must still be in Concepción. Aquino himself, they said, had gone back there on a boat which had left several hours ago.

It was imperative that we found him as soon as possible. I knew only too well, from past experience, that many people merely throw rice or mandioca into the cage of any animal they catch, and if it does not eat they assume it is ill and pay no more attention to it. It might well be that this rare creature was at this very moment starving to death somewhere in Concepción. We must find it and ensure that it was being properly tended, but we had only two days in which to do so, for we could not abandon our Chaco plans.

We dashed over to the airline office. A plane was leaving for Concepción the very next day and there were two spare seats on it. We decided that Sandy and I should take them and that Charles should stay behind to make the final preparations for the Chaco trip.

The plane left Asunción at seven o'clock the next morning, and just over an hour later we landed in Concepción. It was a small quiet town of dusty streets and simple adobe whitewashed buildings. We went straight to its only hotel, for Sandy was sure that this was the best place in which to start our detective work. The patio was crowded with coffee drinkers. I wanted to go from table to table asking if anyone knew a man named Aquino, for we had little time to spare, but Sandy insisted that this would be grossly impolite; many of these people were old friends of his and they would be very offended if he did not greet them in a civilized and leisurely way. One by one, he introduced me to them all. We exchanged polite pleasantries as I tried to contain my impatience.

Sandy explained that I was very interested in giant armadillos. Everyone considered this to be most extraordinary, and said so, at length. Sandy increased their astonishment by revealing that not only was I interested in giant armadillos, but that I actually wanted to acquire a live one. There then followed an extended discussion on the various methods of catching a giant armadillo. This proved to be somewhat unproductive, for no one had ever seen the creature, no one had ever tackled such a job, and everyone made it clear that they had no ambition ever to do so. As a result, the topic veered slightly on to the question of how one would manage to cage the creature once it had been captured. The general consensus of opinion was that this was virtually impossible as it would dig its way out of anything except a steel tank. The waiter sat down beside us and indulged in a little humorous backchat on the subject of what one should offer a giant armadillo to eat and drink. I became increasingly desperate, for after all we had only twenty-four hours to trace it. At last Sandy raised the question of the identity of Aquino. Everyone knew him. He had not yet returned from Asunción, but he usually worked as a lorry driver and had recently been fetching timber from a logging camp run by a German, ninety miles away to the east, close to the Brazilian border. If he had a captive tatu carreta it would undoubtedly be there.

'Can we hire a truck to take us out to the camp?' I asked, realizing, as I spoke, that in all probability this would provoke another half-hour's discussion. Happily, this question was quickly answered, for there was only one man in the whole of Concepción who owned a truck which could do the job. His name was Andreas, and a small boy was dispatched to find him.

While we waited, I visited a nearby shop to buy something on which we might feed the armadillo. All I was able to get were two tins of lambs' tongues and a tin of condensed milk (unsweetened), but at least this would provide a reasonable approximation to the diet which had proved successful with our other armadillos.

Half an hour later, Andreas arrived. He was a young man, with a luxuriant black moustache and oiled hair, wearing an American shirt patterned with vivid floral designs. He ordered a cup of coffee and sat down to discuss the proposition. Three cups later he agreed to take us. All he had to do was to visit his mother, his wife, his brother and his mother-in-law to tell them where he was going, fill up the truck with petrol and then he would be ready to set off. By now, I was beginning to feel that we would never leave the coffee bar, but Andreas was as good as his word and he reappeared with a new and powerful truck only twenty minutes later. Sandy and I squeezed into the cab with him and we roared away, our horn blaring, accompanied by shouts of encouragement from the coffee drinkers and the waiter. All things considered, I felt we had done remarkably well to be on our way within four hours of arriving in the town.

Our speedy progress, however, was not maintained, for Andreas suddenly turned sharp right down a small side turning and drew up outside the local hospital.

He explained that he had spent the previous night drinking

with a sailor who had come up the river from Uruguay. His new friend had made the mistake of inviting one of the girls who was also in the bar to join him in a glass of *caña*, whereupon a man standing next to him suddenly and surprisingly stuck a long knife in the Uruguayan's stomach. Now the sailor was in hospital and Andreas was sure that he would be thirsty, so he was taking a couple of bottles of caña to slip under the sailor's pillow when the nurses were not looking. His visit was a short one, but long enough for me to reflect on the importance of observing local customs.

The road, a wide red earth track cut through the forest, was appallingly rutted and pitted with huge potholes. Most of these hazards Andreas managed to avoid by swerving violently from side to side and only rarely did he slacken speed. Every few miles we passed an encampment of conscripts who, in theory, were supposed to be maintaining the road. However, none of them was actually working and, as Andreas pointed out, it was really rather unreasonable to expect them to do so. A conscript's pay is very small, and, as he receives it whether the road is repaired or not, he is much more profitably occupied doing something else, such as chopping firewood to sell to passing travellers. It did not seem to me that many of them were doing even this, for most were fast asleep in the shade of the trees by the roadside. It was extremely hot and I would have sympathized with them if it had not been for the fact that my teeth were rattling in their sockets and my head continually banging on the roof of the cab as we careered onwards over the increasingly uneven road.

We reached the lumber camp at five o'clock. It was merely a single hut, with some big wheels for carrying logs like those we had seen at Ihrevu-qua standing in front of it. As we approached it, my heart beat uncomfortably fast. Was the giant armadillo alive? With difficulty I prevented myself from running along the track which led to the hut.

The hut was deserted. Not only was there no one at home,

but I could see no sign of an armadillo, nor any place where one might have been caged. However there were signs of occupation – an old shirt, three shining axes, some enamel plates leaning to dry against the log walls, a giant mirror-fronted wardrobe, and an empty hammock slung across one corner. Presumably the German must be out at work in the forest. We hallooed and yodelled. Andreas sounded a braying fanfare on the horn. But no answering sounds came from the forest. We seated ourselves disconsolately in the shade of the hut walls and waited.

At six o'clock, a man on horseback came round a bend in the road ahead. It was the German. I ran towards him.

'Tatu carreta?' I said anxiously.

He looked at me as though I were a raving lunatic. At that moment I realized that we were not going to find a giant armadillo that day.

Sandy extracted the full story and, aided by a little deduction, all became depressingly clear. A week earlier a Pole who worked for the German had come into the camp from a far distant section of the forest where he had been surveying the timber. Over the evening meal he mentioned that he had met an Amerindian who had said that he had recently enjoyed a magnificent feast at his village in which the main dish had been a giant armadillo. The Pole remarked that he had never seen this rare creature and that when he went back he was going to ask the Amerindians if they could catch another to show him. Aquino, who had come up from Concepción to collect a load of timber, overheard the conversation. Obviously he remembered some gossip about the Englishmen in Asunción who were looking for armadillos, and without saying anything to the Pole he had returned to Concepción with his load and had accompanied it down to Asunción. There he had traced the gossip to Sandy's travel agency and, in order to make his bargaining position a better one, had claimed that he had already captured a tatu carreta. Now he must be on his way back and was no doubt going to offer the Pole a trifling sum for one of the animals. Then he would take it down to Asunción and make a vast profit by selling it to us. The German thought this was vastly amusing – not only because we had come so far for a mere animal, but because we had unknowingly thwarted Aquino's plans to make a fortune. He produced a bottle of whisky and passed it round.

'*Musik!*' he cried, and from out of the wardrobe he dragged an enormous piano accordion. Andreas was delighted, and the pair of them began a highly inaccurate version of 'O Sole Mio'. My heart was not in the singing. I was bitterly disappointed. It was gone ten o'clock when we finally persuaded Andreas to start up the lorry again. We left the German with a firm offer to pay well for any giant armadillo that was brought in, detailed instructions on how to care for it, two tins of lambs' tongues and a tin of condensed milk (unsweetened).

By the time we had returned to Asunción, the next day, I had largely recovered from the crushing disappointment of discovering that Aquino's armadillo was a fiction, and as I recounted to Charles the story of our excursion I began to feel a little more optimistic. Although we had not actually set eyes on the beast, we had at least spoken to a man, who employed a logger, who had met an Amerindian, who had eaten one. It was really, I insisted, quite a narrow miss and there was every chance that the rewards I had asked the German to tell the Pole to mention to the Amerindians, would be enough to persuade them that a giant armadillo could be converted into something more valuable than a few pounds of rather tough stewing steak. We might get one yet.

I sensed that Charles remained unconvinced.

27

Ranch in the Chaco

━━━

W e left for the Chaco the day after Sandy and I returned
from Concepción. Early in the morning we piled all our
equipment on to a lorry and drove to the airport. When we
arrived it was immediately clear that there was no chance whatever of packing all our baggage into the tiny plane which was
to take us. We tried, but it was impossible. Something had to be
abandoned. Reluctantly we decided that it should be the food,
for the rancher with whom we were going to stay had insisted,
on the radio, that there was no need to bring any supplies
whatsoever. It was a decision that we were to regret later.

We took off, and as we circled Asunción we looked eastwards
for a moment towards the verdant hilly country, rich with orange
groves and smallholdings, that began just outside the town and
was the home of three-quarters of the inhabitants of Paraguay.
Then we swung west over the Paraguay River, a broad brown
ribbon glinting in the sun, and saw ahead of us the Chaco. From
the very edge of the river, it looked totally different from the
land so close to it on the opposite bank. We could see no signs
of human habitation. A stream wound across it, twisting so
extravagantly that in many places it looped back upon itself; the
current, seeking a more direct course, had cut through the necks
of the meanders so that the forsaken stretches of the river were
left as weed-clogged stagnant lakes. I saw from the map that this
river was very understandably called the Rio Confuso. Here and
there, the land had been colonized by palms which were scattered
thinly over wide areas, like a thousand hatpins stuck in a faded

349

green carpet, but for the most part, there were no houses, no roads, no forests, no lakes, no hills, nothing but a desolate feature-less wilderness. I noticed that our pilot had armed himself with two large pistols and a well-filled ammunition belt. Perhaps, after all, the Chaco was as uncomfortable and as dangerous a place as our acquaintances in Asunción had claimed it to be.

We flew westwards over this ferocious inhospitable country for nearly two hundred miles until at last we sighted Estancia Elsita, our destination.

Faustino Brizuela, the *patron,* and his wife Elsita, after whom he had named his ranch, were waiting for us by the side of the airstrip as we landed. He was a huge man, though his mammoth girth made him appear shorter than his actual height of six feet, dressed in a strikingly unconventional costume comprising an unmatched pair of violently striped pyjamas, a large pith helmet and dark glasses. He welcomed us in Spanish, flashing a wide and predominantly golden smile, and introduced us to Elsita, who stood by his side – a small rotund lady, holding a baby in her arms and chewing an unlit cheroot. A group of half-naked painted Amerindians had also come to meet us. They were tall, barrel-chested men, with straight black hair tied at the back of their heads in ponytails. Some held bows and arrows, one or two carried antiquated shotguns. In the weeks to come Faustino seldom appeared in public without his pyjamas or Elsita without her cheroot, but the appearance of the Amerindians was not typical; they had decked themselves specially for our arrival, and never again did we see them looking so spectacular.

An acquaintance in Asunción had told us that the ranchers of the Chaco were lazy people, and to prove it he related the story of an agricultural expert from the United Nations who, visiting an estancia deep in the Chaco, had been appalled to find that the rancher lived on nothing but mandioca and beef.

'Why don't you grow bananas?' the expert asked.

'Bananas just don't seem to grow here; I don't know why.'

'What about pawpaw?'

'That doesn't seem to grow either.'

'And maize?'

'It just doesn't grow.'

'And oranges?'

'Same trouble.'

'But a few miles away there is a German settler and he grows bananas, pawpaws, maize and oranges.'

'Ah yes,' replied the settler, 'but he *planted* them.'

If Faustino was typical, however, the story was unjust, for the central courtyard of his house was shaded by orange trees, laden with ripe juicy fruit, pawpaws grew by the kitchen door and beyond the garden stretched an acre of tall betasselled maize. Above the red pantiled roof, an aluminium-vaned windmill spun in the breeze, generating electricity for lighting the house and running the radio. Furthermore, Faustino had even devised a method of supplying the kitchen and bathroom with running water. By the side of a large, duckweed-covered lagoon which lay near the house he had dug a shallow well, lining its sides with wooden boards. Above it he had built a scaffold which supported a large iron tank and this was filled every morning with bucketfuls of well water hauled up to it by a rope and pulley operated by a little Amerindian boy on a horse. From the tank, the water ran down pipes to the taps in the house. It was an admirable and very efficient arrangement, and we assumed that the water itself was good, for Faustino, Elsita and their children all drank it freely. It was not until we ourselves had also been drinking it for several days that we had any cause to examine the well in detail.

We needed some frogs to feed a cariama, a large bird which one of the peones had brought in to us, and Faustino suggested that we should find an endless supply of them in the well. I went down to it and swirled my net in the turbid, slightly smelly water. When I took it out, I found that I had caught three lively olive-green frogs, four dead ones and a decomposing rat. Maybe the rat had accidentally fallen in and drowned, but what ingredient of our drinking water had killed such accomplished swimmers as the frogs was a zoological problem that I did not care to investigate. For two days afterwards, we surreptitiously dropped chlorine tablets into any water that we drank, but they produced such a revolting taste that eventually we abandoned the habit.

———

We had arrived at the end of the dry season. Most of the *esteros*, once gigantic swamps, were now barren tracts of baked mud, frosted by salt which had been precipitated from the evaporating waters, hummocked by the root clumps of withered reeds, and pockmarked by the deep rock-hard footprints of the cattle which months ago had plodded across the swamps to reach the last puddles of water. In the centre of some of these there still lingered patches of glutinous blue mud in which our horses sank up to their hocks. In a few places we found shallow lagoons of muddy tepid water like that which lay close to the house, the last remnants of the annual floods which had so recently covered the greater part of the country.

Only where the ground rose slightly above the general level of the surrounding country was it possible for trees or bushes to grow, safe from death by drowning, and such areas had been colonized by scrub vegetation, the *monte*. All the plants bore savage spines which protected them from the grazing cattle, desperate for fodder in the drought, and many had also developed devices to enable them to conserve water during the dry season. Some did so in their huge underground roots, others, like the hundred-armed candelabra-like cactus, in their swollen fleshy stems. The *palo borracho*, the drunken tree, conserved moisture in its distended bloated trunk, thickly studded with conical spines. These trees, perhaps, epitomized the character of the armoured vegetation of the Chaco, standing in groups like grotesque bottles which had come to life and sprouted branches.

———

The Amerindians lived half a mile from the estancia house. Not many years ago, these people, the Maká, were considered untrustworthy and murderous and no doubt the early pioneers who invaded their country had given them every cause to be so. Originally they seldom stayed long in one place but wandered over the Chaco, building temporary encampments wherever game was relatively plentiful. Most of the people in this village, however, had abandoned their traditional hunting life and many of the men worked as peones on Faustino's estancia. Their tolderia was, in fact, a permanent settlement, but even so, the style of their houses had not changed or elaborated – they remained simple dome-shaped huts, roughly thatched with dry grass. The people spoke a language quite unlike any that I had heard before. It consisted predominantly of guttural words each of which, as far as I could tell, had the emphasis placed on the last syllable so that their speech sounded remarkably like a tape recording of an English voice played backwards.

On our first afternoon we were met by Spika, one of the Amerindians, who followed us as we wandered among the huts. Suddenly I came to a halt. In front of me, hanging from the rafters of a rough shelter built over a fire, I saw a basket made from the shiny grey shell of a nine-banded armadillo.

'*Tatu!*' I said excitedly.

Spika nodded. '*Tatu hu.*'

Hu, in Guarani, means black.

'*Mucho, mucho?*' I asked, waving my arm at the surrounding countryside.

Spika was quick to grasp my meaning and nodded again. Then he added something in Maká which I did not understand. I looked puzzled and to explain what he meant Spika picked up from among the ashes of the fire a fragment of shell and handed it to me. Although its broken edges were scorched and blackened, enough of it was undamaged for me to recognize it as part of the yellow tessellated shell of a three-banded armadillo.

353

'*Tatu naranje*,' said Spika. '*Portiju*,' he added, licking his lips in an exaggerated mime of a hungry man.

This Guarani word I had already learned from Faustino. It means roughly, 'Excellent food.'

By a further combination of Spanish, Guarani and gesture, Spika explained that tatu naranje, the orange armadillo, were quite abundant in and around the monte; that although they came out at night, they could also be found during the daytime; and that it was not necessary to build traps for them, because once you found them you could pick them up by hand.

He also told us that there was another type of armadillo to be found in the neighbourhood, the *tatu podju*. *Podju*, Sandy told us, meant 'yellow-pawed', but from this meagre description, I was unable to be sure what kind it was. Nonetheless, we had discovered that there were at least two species of armadillo to be found hereabouts that we had not seen before, and the next day we borrowed horses from Faustino and set off to try to find them. Candidly, in spite of what Spika had said, I found it hard to believe that we should see them during the daytime, but at least we should be familiarizing ourselves with the lie of the land so that, if it became necessary to hunt at night, we should be able to do so without getting lost.

But Spika was right. We were not more than a mile or so from the house when we saw an armadillo crossing a dried-up marsh, an *estero*, only a few yards ahead of us. While Sandy held the reins of my horse, I set off on foot in pursuit. The armadillo was over two feet long – considerably bigger than a tatu hu, his yellowish-pink shell was sparsely covered with long bristly hairs and his legs were so short that I found it hard to believe that he could run very much faster even if he wanted to. So instead of catching him immediately, I trotted by his side to see what he would do. He stopped for a moment to look up at me with his tiny bewhiskered eyes and then he trundled on over the rough surface of the estero, grunting loudly to himself. Soon he came across a depression in the ground. He sniffed it and began to dig, throwing back great quantities of earth with his forepaws. Within a few seconds, only his hindlegs and tail were showing and I decided that the time had come to catch him. With his head buried in the hole, he was unaware of my intentions and unable to take avoiding action, so that all I had to do was to seize hold of his tail and gently extract him. He came out, puffing and grunting, and still making breast-stroke movements with his forelegs.

We took him back to the house and Spika came along to identify him.

'Tatu podju,' he said approvingly, and 'Podju' thereafter became his name. Scientifically, he was a six-banded, or hairy, armadillo. In Argentina, this species is called the *peludo*. Hudson was full of admiration for these animals which he considered to be the most adaptable in diet and habit of all the creatures of the pampas. He tells one extraordinary story about the way in which one made a meal of a snake. The peludo crawled over the angry hissing reptile and swayed backwards and forwards so that the jagged edge of its shell lacerated the snake and almost sawed it in two. Again and again the snake struck at its attacker, but without effect and eventually it died, whereupon the armadillo began to eat it, tail first.

Every day we explored the surrounding country. Sometimes we went out on horseback with Faustino or the peones. Admiringly I tried to emulate their manner of riding, so different from the bobbing style practised in England, for they sat fast in their sheepskin-padded saddles as though they and their mounts were welded together. We started by wearing all the equipment that we had purchased in Asunción – bombachos, riding boots, leather leggings and fajas. One by one we discarded them. The voluminous floppy bombachos, though excellent and cool when riding, were serious handicaps when we dismounted and walked into the thorny monte. My boots, after a single excursion into a bog, dried in a distorted shape and became so excruciatingly uncomfortable thereafter that I could not bear to wear them. The leather leggings were far too hot and stiff, and the faja, though decorative and highly professional-looking, had to be worn so tightly if it were to serve its proper function that I preferred to abandon it and run the risk of my 'guts banging about'. Only the ponchos were of any real value to us – we used them as padding for our saddles.

At other times we went on foot. The nearest tract of monte began just beyond the tolderia and stretched for several miles northwards to the banks of a salty sluggish stream, the Rio Monte Lindo. In its denser parts, it was virtually impenetrable. Giant cactus, thorn bushes and stunted palm trees were matted together with lianas; the ground was overgrown with the fleshy rosettes of caraguata; and every plant, bush and tree bristled with spines, daggers and barbs that snatched at our clothes, stabbed through our canvas shoes and ripped our flesh.

Here and there *palo santo* and *quebracho* trees grew high above the thorn thickets and in a few places the bushes thinned into desolate meadows of isolated cacti growing among tussocks of coarse grass.

Some of the birds which lived here seemed to be possessed by a mania for nest-building and constructed mansions so large that they were extremely conspicuous. In one clearing, we found a dozen stunted thorn trees, each of which supported in its topmost branches, untidy ricks of twigs nearly twice the size of a football. The architects of these homes, small drab birds a little smaller than thrushes, perched on top of their nests singing shrilly in the ferocious sun. Sandy called them *Leñatero* – Firewood Gatherers. Some of them were still busy building. Although they were not powerful fliers they were nonetheless extremely optimistic about what they could carry and selected twigs of a size and weight that would have daunted many a larger bird. We watched them fly valiantly on rapidly beating wings towards their nests, carrying cross-ways in their beaks twigs that were longer than they were. As they approached their nests, they sometimes were unable to summon up the necessary strength to enable them to make a clean landing and their twig would catch on part of the bush. As a result, piles of abandoned sticks and twigs had accumulated beneath each nest. This would have made excellent kindling wood for a camp fire and it was easy to see how the leñatero had acquired its name.

The largest of all nests, we found in a dead palo santo tree which stood alone just beyond the margin of the monte, gaunt and barkless, its naked trunk bleached by the sun. Around its branches had been built several elongated constructions of twigs and sticks, the size of large corn stooks. These were the homes of a colony of Quaker parakeets, green birds with grey cheeks and undersides, about twice the size of budgerigars. All other members of the parrot family nest in holes of some sort – in tree trunks, in the globular nests of termites and tree-ants, or in burrows. The Quakers alone build in the open. Their huge nests were not communal homes but rather more akin to blocks of flats, for each pair of birds possessed its own separate nesting chamber, porch and entrance, and there were no tunnels or passages which connected one chamber with another.

They were extremely industrious creatures. Parties of them were continually arriving with fresh supplies of green twigs which they had cut from bushes in the monte and those that had stayed at home were busy filching material from any of their neighbours' territories that happened to be unguarded. This feverish building activity never stops, for the Quakers live in their nests throughout the year. Renovations are necessary before the breeding season, nurseries have to be enlarged to cater for the growing youngsters, and the young birds, when they grow up, often build their own accommodation close to their parents' home. So the colony grows larger and larger until it is in real danger of being blown down in a gale.

Quaker parakeet nests

Even the most unobservant traveller could not fail to notice the conspicuous nests of the Quakers and the leñateros, but not all the Chaco birds built so boldly. One day I took an exploratory walk along a hunting trail that the Amerindians had made through the monte. After an hour or so, streaming sweat, my mouth parched, I sat down in the meagre shade of a thorn bush and gulped from my water bottle. As I wondered whether or not I should turn back, I heard a buzzing noise above me. I looked up and saw a tiny green hummingbird hovering among the branches. I could not imagine what had brought it there, for there were no flowers in this tree from which it could sip nectar. It was, however, busily occupied with something, darting to and fro among the branches, and hovering on its wings which were moving so fast that they were merely a vague blur. Hummingbirds are capable of beating their wings at the almost unbelievable speed of two hundred times a second, but they

358

only do so when diving, or indulging in courtship flights. The little creature above me probably did not need to make much more than fifty beats a second to enable it to hover and it was only when it accelerated to dart away into another position that it increased the rate of its wing-beat enough to produce the humming noise which had first drawn my attention. Suddenly it was gone, darting across the clearing in front of me like an arrow.

Most hummingbirds are polygamous and each female builds a nest for herself, taking total responsibility for incubation and feeding her nestlings. I knew therefore that I was watching a hen bird. With her scarlet bill, she spread the spiders' silk she had collected around the outside of the tiny nest. When it was all in place, she began darting out her thread-like tongue, producing a sticky spittle which she smeared around the outer surface of the nest, using her bill as though it were a palette knife and she icing a cake. Then she began pedalling furiously with her feet, revolving on the nest as she did so, to shape and smooth the inside of her cup. After a few more strokes of her bill she once again darted away to gather another batch of material.

She worked so hard that after an hour I was sure that the nest had visibly increased in size since I first saw it. I had sat so long in silence studying the hummingbird that the other creatures of the monte seemed no longer to be aware of my presence. Small lizards pattered around on the bare earth between the grass tussocks; a working party of Quaker parakeets settled on a thorn bush and began to collect building materials, squawking and chattering among themselves. As I watched all this activity around me I thought I saw, out of the corner of my eye, a slight movement beneath a clump of spiny cactus. I searched the area with my binoculars but I could see nothing among the withered grass and the twisted fleshy stems of the cactus except a round clod of yellowish earth. And then, as I watched, the clod moved. A dark vertical line appeared in its lower half which slowly expanded. Then with a jerk a little hairy face peered out and the ball was transformed into a tiny armadillo. It was a tatu naranje. Gingerly it pushed its way through the grass but, as soon as it reached an open space, it increased its speed, running on the tips of its toes, its little legs moving with such rapidity that it looked like some odd clockwork toy. I jumped up and set off in pursuit. The armadillo executed a neat swerve and disappeared down a low tunnel beneath the leaves of a patch of caraguata. I hopped over the plant and waited for the armadillo to emerge, feeling as though I were playing trains. In a few seconds out came the armadillo straight into my hands.

The little naranje grunted angrily and snapped tight, transforming itself once more into a yellow ball, its scaly tail fitting beside the horny triangular shield on top of its head, so that no unarmoured part of the body was exposed. In this position nothing could harm it except perhaps a wolf or a jaguar which might be able to crack it open with their powerful jaws. I took a cloth bag out of my pocket and put the rolled-up naranje inside. These bags are extremely useful for carrying newly caught animals of all kinds. Being loosely woven they allow enough air to pass through to enable the creature to breathe properly and, in the darkness, animals nearly always lie still and do not struggle and injure themselves. I put the bag with the armadillo inside on the ground and went back to my place by the hummingbird's nest where I had left my binoculars case. When I came back, the bag had gone. I looked around and saw it, moving slowly along the ground turning over and over. The little clockwork naranje had uncurled and was running away inside the bag. I gathered it up and took it back to join Podju in the semi-derelict oxcart in which he was happily confined.

Within a week, we had assembled three pairs of naranje and two pairs of nine-banded armadillos as well as Podju. The oxcart was fully big enough to house them all, but the amount of food they ate was enormous and we put in such quantities each evening that we began to refer to it as the Soup Kitchen.

The tatu naranje – open and (below) curled into a tight ball

There was no shortage of beef at the estancia, for a cow was slaughtered every week. But meat in itself was not sufficient. The armadillos needed milk and eggs as well, and these were not so plentiful. Fortunately, by now, one of the hens which passed in and out of our bedroom had decided to make its nest in my kitbag. My initial reaction had been to turn her out, but as she produced an egg each day I deceitfully said nothing to Faustino and Elsita and added it to the armadillos' feed each evening together with as much milk as could be spared for us from the kitchen.

The naranjes, however, did not settle down well. The tender pink soles of their feet began to develop raw patches. To prevent this we lined the bottom of the Soup Kitchen with earth. This cured the trouble but involved us in a great deal of extra work, for the armadillos were such messy feeders that they spilt much of their food which tended to putrefy in the earth and turn it sour. We had, therefore, to clean it out every few days and put in new soil.

Then the naranjes began to develop severe diarrhoea. It was only too easy to discover which of them were afflicted for they were very highly strung little creatures and when we picked them up, not only did their legs quiver in alarm, but they always obligingly produced samples of their droppings. We tried varying the proportions of their food. We experimented by adding boiled mashed mandioca to it – but they refused to eat it. The diarrhoea got worse. Both Charles and I became very worried. If we could not cure them, we felt we must turn them loose rather than let them die in captivity. We discussed the problem endlessly. Then it occurred to us that in the wild, the naranjes, grubbing about for insects and roots, would inevitably consume a great deal of earth. Perhaps their digestions required it; perhaps the food we were offering them was too rich. That evening we added two handfuls of soil to our mixture of minced meat, milk and eggs and stirred it up into an unattractive runny mud. Within three days the naranjes were cured.

28

Chaco Journey

When the wind blew from the south, it brought with it chillingly cold weather and often hours of drenching depressing rain. On such days of forced inactivity we often visited the open-sided thatched hut by the corral where the peones congregated to chat, to sharpen their knives and plait rawhide lassoes, to flirt with the half-Indian girls from the estancia kitchen, and – most important of all – to drink hot maté. A woodfire usually burnt in the centre of the floor and the peones would always make room for us on the bench so that we might warm ourselves by the flames and share in the maté as it passed from hand to hand. It was a friendly hospitable place, smelling of horses, leather and the fragrant smoke of palo santo wood.

One rainy morning, when I went up to the hut in search of a drink of warming maté, I found to my disappointment that the place was deserted, except for half a dozen sleek and well-fed dogs. As I arrived they sat up and looked at me suspiciously. Then I noticed a man stretched full-length on his back on the wooden bench, with a dusty broad-brimmed hat flat over his face. As far as I could tell, I had never seen him before. He was very tall – certainly over six feet – and was wearing torn baggy bombachos, an unbuttoned shirt and a faded Amerindian-woven faja around his waist. His feet were bare and it was clear from their horny and calloused soles that he seldom wore shoes.

'*Buenas dias*,' I said.

'*Buenas dias*,' replied the stranger in muffled tones from beneath his hat.

'Have you come far?' I asked in halting Spanish.

'Yes,' he replied, without making any movement except to scratch his stomach lazily.

There was a pause.

'It is cold,' I said, rather pointlessly, but I could think of no topic other than the weather with which to prolong the conversation. The stranger swung his legs to the ground, pushed his hat on to the back of his head and sat up.

He was a handsome man with tightly curling black hair, greying in places, a deeply tanned face and several days' growth of grizzled stubble on his chin.

'Would you care for some maté?' he asked and without waiting for a reply he began undoing the canvas bag which he had been using as a pillow. He extracted a cow's-horn cup, a silver *bombilla* and a small packet from which he poured some dry green maté into the horn. In silence, he added water from the earthenware jar that stood by the bench and sucked at the bombilla. He spat out the first few muddy mouthfuls, refilled it and courteously passed it to me.

'What do you do here?' he inquired.

'We are looking for animals.'

'What sort?'

'Tatus,' I replied airily. 'All kinds of tatus.'

'I have a tatu carreta,' he answered.

At least that was what I thought he said, but I was not certain. Perhaps he had spoken in the past tense; or had said he could catch a tatu carreta if he wanted to. I could not be sure.

'Momentito,' I said excitedly and bolted out of the hut and into the rain to fetch Sandy from the house. When we returned together, Sandy embarked on the polite protracted small talk which he insisted was the correct way to preface any serious inquiry. I sat by, fidgeting with impatience. After a few minutes, Sandy translated a brief summary of his conversation. The stranger's name was Comelli. He was a hunter who roamed over the Chaco looking for jaguar, nutria and fox or anything which

had a skin sufficiently valuable to trade for matches, cartridges and knives and the few other things that he required to enable him to follow his wandering life. He had not slept in a house for ten years and had no wish to do so.

'And the tatu carreta?' I asked anxiously.

'Ah!' said Sandy, as though he had forgotten all about it.

Once again he and Comelli chatted.

'He once had a tatu carreta and kept it for several weeks, but that was a long time ago.'

'What happened to it?'

'It died.'

'Where did he catch it?'

'Many leagues from here, beyond the Pilcomayo River.'

'Could he take us there tomorrow?'

Sandy translated the question. The stranger grinned broadly.

'With pleasure.'

Excitedly I ran back to the house to tell Charles the news. I was keen to leave immediately for the place Comelli described. Whether we found a giant armadillo or not, we might well see the other animals which did not occur around the estancia. It would take us three days to ride there and, if we were to allow any time at all for hunting, we should have to be away for at least a fortnight. Faustino offered to lend us two horses, a cart for our equipment and a pair of oxen to pull it. But we had no stores.

'Oh, we'll live off the country,' I said to Charles enthusiastically, but rather vaguely.

'Well that couldn't be much worse than our food at present,' he replied morosely.

In this, I had to agree with him. Faustino and Elsita were extremely hospitable, but their meals were scarcely appetizing to anyone unused to them, for they consisted exclusively of various parts of the anatomy of a cow – fried intestines, numerous shrivelled but curiously shaped organs which I was, perhaps fortunately, unable to identify, and interminable slabs of leathery meat the texture of vulcanized rubber. It would be a positive relief to 'live off the country' if this implied a change of diet.

We discussed the problem with Faustino.

'The Chaco is hungry country,' he said. 'We can give you mandioca and farinha and maté, but a man will not get fat on that.' (Farinha is dried grated mandioca.)

Then he brightened.

'Never mind. When you get hungry, I give you my permission to kill a cow.'

It took us two days to make all the arrangements. The leather harness of the cart had to be repaired; the oxen and the horses had to be found and rounded up. Elsita looked out the stores and produced a big cast-iron cooking pot and a frying pan. Charles and I gathered a box full of oranges, and Faustino solicitously gave us the left hindleg of a cow which, he explained, would provide us with at least one meal before it went bad in the heat.

At last, all was ready. The cart was piled high with equipment, and the oxen yoked to it. Sandy took the reins and, to the accompaniment of piercing squeaks from the greaseless wheels, we lumbered away from the estancia. The wind had changed to the north, the cold rainy weather had disappeared and we rode beneath a cloudless harshly blue sky. Comelli went ahead, leading the way. With his broad-brimmed hat, his long legs dangling stirrupless and nearly touching the ground, he looked like some South American Don Quixote. His dogs ranged far and wide around us. Comelli knew each of them not only by their voices but by their footprints and, as we travelled, he called them from time to time. The head of the pack he had named Diablo, the Devil; the second in command – Capitaz, the Foreman; there were two others whose names I never learned and a big brown bitch, the laziest and most handsome of them all who was devoted to Comelli and he to her. He called her Cuarenta, because, he said affectionately, her feet were so large that she would take size forty in boots.

We headed south. Soon the estancia and its nearby monte had dwindled to nothing and vanished. Before us stretched the wide flat plain populated only by a few of Faustino's cattle. The oxen plodded slowly forward. They were incapable of walking faster than about two miles an hour and in order to keep them moving at all the driver had to shout at them almost continuously. Having only a couple of horses between the four of us we took turns in riding and in driving the oxen, and when we were doing neither we sat on the tailboard of the cart, sipping cold maté.

In the late afternoon, we sighted on the horizon the gaunt skeleton of a tree. As we drew near, we saw that its top branch supported a huge nest of a jabiru stork. A lake lay in front of it and thorn bushes were clumped around its base.

There we made our first camp.

Comelli and Cuarenta

For the next three days we made our way southwards across the plains. Comelli spoke of the copses of monte as islands, and the name was apt – they were islands of bush in a sea of grass and, like a man at sea, Comelli used them as landmarks by which to navigate. The weather since we had left the estancia had been suffocatingly hot and the sun had scorched us as we rode, but on the morning of the fourth day, the wind changed, the sky clouded over and it was raining hard by the time we reached the Rio Pilcomayo in the late afternoon.

The river itself had divided into several streams which slipped muddily between untidy spits of gravel. Eighty years ago the Pilcomayo had been accepted as the boundary between Argentina and Paraguay but since that time the river had changed its course countless times as it wandered across the flat Chaco. Now it flowed many miles north of the channel it had followed when the frontier had been agreed, so that the land on the southern side was still in Paraguayan territory.

We urged our horses into the river. Though it was not deep, the water swilled perilously close to the floor boards of the cart before the oxen finally hauled it on to the other bank.

Two days previously, we had finished the beef Faustino had given us. We had found no game and already our diet of unrelieved mandioca, farinha and maté was beginning to pall. Comelli, however, assured us that we were heading for a small trading store, called Paso Roja, which was always stocked with tinned foods of every description. My mouth watered at the thought.

We reached the tract of monte which sheltered the store in the late afternoon during a downpour. The equipment was in danger of getting soaked unless we could find some shelter and Comelli led us along a muddy track through the thorn scrub to a small derelict shanty. It consisted of nothing more than four crumbling adobe walls and a sagging thatched roof. Comelli told us that the man who had built it had died there a few years earlier and now lay buried somewhere in the monte. Since then the hut had been deserted. The rain gushed from its roof, a wide pool lay on the threshold and the wind whistled through the gaps in the walls. Hastily, we unloaded the cart and stacked the equipment in the few places inside the hut which were clear of the drips falling from the leaky roof.

We were all tired, wet and hungry, and as soon as we had finished we went together through the rain to the store which stood half a mile away in the monte. It was a little larger than the hut we had appropriated, but almost as dilapidated. We walked through the open doorway, stepping between the bedraggled chickens and ducks which had collected inside to shelter from the storm. Two hammocks were slung across the room and in one of them lay the *patron*, drinking maté. He was surprisingly young and, to my mind, unaccountably cheerful. As we introduced ourselves, he called his wife and another young man, his cousin, from the back room to meet us. We sat ourselves on some wooden boxes, shivering in our wet clothes, and Sandy asked if we could buy some food.

Patron smiled happily and shook his head.

'None,' he said. 'I've been expecting an oxcart to come up with supplies for several weeks, but it hasn't arrived. All I have is beer.'

He went into a side room and brought out a crate containing six bottles. One by one, he handed them to his cousin who proceeded, to my alarm, to remove the metal tops with his back teeth.

We drank from the bottles. The beer was thin and cold, the last sort of refreshment I would have chosen for preference and certainly no substitute for the tinned sardines and peaches that I had been mentally relishing for the whole day.

'*Paso Roja*, very good?' asked Comelli jovially, clapping me on the shoulder.

I gave him a wan smile of assent but I could not bring myself to express the lie in words.

That night we lit a fire in our hut so that we could dry our soaked clothing and cook an unappetizing meal of farinha. There was not enough room for all four of us and the dogs to sleep under cover so Charles and I volunteered to spend the night outside, for although it was still raining hard our hammocks, which had originally been made for American troops serving in the tropics, were fitted with a slim rubberized roof and were, in theory, waterproof.

Not far from the hut stood a ruined outhouse. Its roof and three of its walls had fallen, but its corner posts were still standing. During a lull in the storm, I ran out and slung my hammock between two of them. Charles tied his between two tall trees nearby. Within a few minutes, I was inside mine and out of the rain. I zipped up the mosquito net that joined the roof to the main part of the hammock, wrapped myself in my poncho, put my torch to my side and went to sleep almost immediately, feeling warmer and more comfortable than I had all day.

I awoke soon after midnight with the unpleasant sensation that my feet were approaching my head and that I was folding up like a jackknife. I fumbled for my torch and by its light discovered that the posts from which I was suspended were leaning drunkenly towards one another and that the hammock had sagged to within a few inches of the ground. I lay still and considered the situation. The rain was still falling heavily, puddling the earth around me. If I clambered out, I should be soaked to the skin within a few seconds. Yet if I stayed in my hammock, the posts would slowly hinge closer together until I was deposited on the ground. I reasoned that when I reached that position, I should be no worse off than sleeping on the floor of the hut so I decided to stay where I was and went back to sleep.

I was woken an hour or so later by a cold wet feeling in the small of my back. I did not need my torch to realize that I was virtually sleeping on the ground and that I had settled in the middle of a large puddle which was slowly seeping through both my hammock and poncho. I lay in this position for half an hour, watching the lightning illuminate the driving sheets of rain. I tried to weigh the discomfort of my present situation and balance it against the certain soaking I should get if I returned to the hut. The thought of warming my chilled body by the embers of the fire in the hut finally tipped the scales. I unzipped the mosquito net, abandoned my sodden bed in its puddle, and splashed in my bare feet across the muddy clearing.

The hut was reverberating with the snores of Sandy and Comelli and smelt strongly of wet dogs. The fire was out. Miserable and cold I squatted in a vacant corner. Cuarenta had noticed my arrival and stepped delicately over Sandy's outstretched legs to settle herself at my feet. I wrapped my wet poncho around me and awaited the dawn.

Comelli was the first to wake. Together he and I blew life into the fire and put on a saucepan of water to boil for maté.

With the coming of dawn, the storm cleared. Charles awoke in his hammock, stretched himself luxuriously, announced that he had spent a splendid and most comfortable night, and remarked facetiously that he would like his maté in bed that morning.

I did not consider his joke to be in the best of taste.

During breakfast, Patron, Bottle-opener and a third man strolled into the hut and seated themselves by the fire. Patron introduced the stranger as another of his cousins whose chosen vocation was to slaughter cattle. He was a surly-looking man and his appearance was not enhanced by a puckered scar which stretched across his face, twisting his eyebrow, contorting his eyelid and pulling the side of his mouth into a leer. Patron explained that the scar had been inflicted during an evening of heavy drinking when Slaughterer had been irritated by Bottle-opener and had set on him with his butchering knife. Bottle-opener defended himself with a broken caña bottle with such effect that Slaughterer sobered up very quickly and Patron's wife was summoned to sew up his wound. The three cousins, however, still seemed to be the best of friends, which was as well, for they were the only inhabitants of Paso Roja and there was no other homestead for many miles.

We told them that we were looking for animals of all sorts and were particularly interested in tatu carreta. Bottle-opener agreed that he had once found the spoor of one, but none of the men had ever seen the animal itself. They promised us that they would keep their eyes open for any creatures which might interest us.

It was clear that they intended spending the morning with us. First they asked to see our equipment. Bottle-opener was captivated by Charles's hammock and climbed inside where he remained, lost in admiration of the zips, the mosquito net, the pockets and roof. Patron sat outside the hut on a log inspecting my binoculars with reverence, turning them over and over, stroking the barrels and occasionally putting them to his eyes. Slaughterer was professionally interested in knives. He found mine and squatted by the fire admiringly testing its blade with his thumb and dropping very obvious hints that he would welcome it as a gift. As I did not respond to them – it was after all the only small knife I possessed – he changed his line of approach.

'How much?'

'One tatu carreta,' I replied without hesitation.

'The bitch!' he said somewhat enigmatically, using a rather vulgar Spanish word, and with a flash of his arm, he threw the knife so that it stuck quivering in the trunk of a tree fifteen feet away.

After breakfast, Comelli suggested that he should leave on a long tour of the monte to the east to see if he could find any traces of giant armadillos. He would need two or three days to cover all the territory that he wanted to inspect, but if he found anything interesting he would return immediately and fetch us. It took him no more than a few minutes to gather together the few things he required – a poncho, a bag of farinha and another of maté – and before the sun was above the trees, he rode quietly away on one of the horses, his dogs trotting ahead of him, their tails waving happily.

Sandy volunteered to spend the day repairing the hut, building a shelter to serve as a kitchen and generally endeavouring to impose some order on our untidy and hastily established camp.

As we now had only one horse between us it was not possible for Charles and me to go on a long journey together, so instead we decided to load the remaining horse with our cameras, the recording machine and water bottles, and set off on foot to reconnoitre the plains to the north.

The Chaco between Paso Roja and the Pilcomayo was diversified by *riachos* – long shallow creeks, some nearly a hundred yards long, which arose from nowhere and ended abruptly and unexpectedly in a puddle of mud. They were filled with water hyacinth and other floating weeds, and hummed with mosquitoes and huge vicious horse-flies. On the banks of one of them I discovered an interesting-looking pile of dried reeds. I poked among them cautiously with my bush knife and found in the damp lower layers a dozen little caiman, the South American relatives of the crocodile. Several of them scuttled over my feet and flopped into the water of the riacho, but I managed to pick up four. The reeds were the remains of a nest in which the mother caiman had laid her eggs and then left them to hatch in the heat of the sun. The ones I had caught were babies, only about six inches long, but young though they were they snapped at my fingers, made angry honking noises and glared at me ferociously with their jaws apart so that they exposed the lemon-yellow leathery lining of their mouths. I dampened a cloth bag and dropped the little reptiles inside.

As we looked round to see what else might be living in the riacho, I suddenly realized that on the opposite bank four men were standing watching us in silence. They were Amerindians. Each of them carried a long ancient-looking gun. They were naked to the waist, barefooted and wore only trousers and leather leggings. Their faces were tattooed and their long hair hung in

373

matted locks down their cheeks. Two of them carried bulging bags and one was holding the butchered, plucked carcass of a rhea.

Obviously they were hunters. Here was an opportunity to recruit highly skilled assistants. We paddled across the riacho to join them and tried to suggest by gestures that they should return to our camp with us. They listened to me with puzzled expressions on their faces, leaning on their guns. At last I succeeded in making my meaning clear and after a rapid guttural conversation among themselves, they nodded agreement.

Back at camp, Sandy was able to talk to them in Guarani and Maká. He discovered that they had left their tolderia many days before and had been hunting for rheas. Rhea feathers fetch a good price, for in Argentina they are used to make dusters. Consequently the Amerindians are easily able to sell them to traders such as Patron and obtain matches, salt and cartridges in exchange. They had disposed of their feathers somewhere close to the Argentine border and were now on their way back to their village. Sandy explained to them that if they would help us in our search for animals, we would supply them with farinha for as long as they stayed with us and that we would pay good rewards in addition for every animal that they brought to us. We would give a specially high price for a tatu carreta. They agreed to join us and immediately demanded some maté as an advance payment. We gave them several cupfuls from our dwindling supply. I hoped that, as the terms had now been agreed, they would stride off into the monte and begin hunting straightaway, but they interpreted their duties somewhat differently. They lay down in the shade of the trees, pulled their ponchos over their faces and went to sleep. Perhaps, after all, it was a little late to begin searching that day.

They awoke at sundown, built a fire some distance from our own and came over to ask us for some farinha. We supplied it, and they took it back to their fire where they began to boil it with joints of rhea meat. The moon rose, large and silver above the trees, and we began to prepare for sleep. The Amerindians, however, seemed refreshed by their afternoon's siesta and showed no signs of retiring early.

'Perhaps,' I remarked hopefully to Sandy, 'they are planning to hunt all night.'

Sandy laughed mirthlessly. 'I'm sure they are planning no such thing,' he said. 'But there is little point in trying to persuade them. You can't hurry an Indian.'

I re-rigged my hammock between two trees, climbed in and composed myself for sleep. The Amerindians seemed very happy, shouting and laughing among themselves. From where I lay, I could see that they had a bottle of caña which they were passing from one to another, taking long swigs from it. The party grew more riotous and with a whoop one of them threw the now empty bottle into the bushes beyond the fire. I watched one of his companions fumble in his bag and produce another full bottle. It would be a long time before they settled down for the night. I turned over, pulled my poncho over my head and tried once more to go to sleep.

Suddenly there was a deafening explosion and something hummed over my head. I peered out in alarm and saw that they were now cavorting around their fire, one of them holding a caña bottle and all of them brandishing their guns. One of them let out a yell and fired his gun into the air again.

The situation was now getting out of hand and something had to be done before someone was injured.

Charles was already out of his hammock and in the hut. I joined him and found that he was urgently unpacking the medical kit.

'My God!' I said. 'Did they hit you?'

'No,' he replied darkly, 'but I'm going to make quite sure that they don't.'

One of the men lurched over towards us, mournfully holding a caña bottle upside down to show us that it was empty. He mumbled something which seemed to be a request for a refill.

Charles handed him a mugful of water and dropped something into it.

'A sleeping pill,' he said to me. 'It can't hurt him and with any luck, we won't need to dodge more than a couple of volleys before it works.'

The others gathered round, and stood swaying unsteadily, anxious not to miss whatever it was their companion had been given. Charles obligingly supplied each of them with a pill. They all threw down their dose in a single gulp and blinked, surprised that the drinks we had given them had no appreciable taste.

I had no idea that sleeping tablets could take effect so quickly. The first man dropped his gun and sat down heavily, shaking his head. For a few minutes he tried to sit up, nodding groggily, until finally he collapsed on his back. Soon all four of them were fast asleep. At last there was silence in the camp.

In the morning the men lay exactly where they had collapsed the previous night. None of them stirred until mid-afternoon. The caña had left them with the most dreadful hangovers and they sat miserably beneath the trees, bleary-eyed, their hair hanging untidily over their faces.

That afternoon, they drifted out of the camp. I had hoped maybe, that the party being over, they had now decided to begin work in order to pay off what they owed us in maté and farinha; but we never saw them again.

29

A Second Search

———

Cuarenta trotted into the camp two days later and greeted us with an effusion of barks and licks. Diablo stalked in behind her, dignified and aloof, leading the rest of the pack, and the dogs lay down together beneath one of the trees. Ten minutes later, Comelli appeared round the corner of the bend, jogging easily on his horse, his hat on the back of his head and a great rent in his bombachos. As soon as he saw us, he shook his head sadly.

'Nothing,' he said, dismounting. 'I went many miles east as far as the end of the monte, but I found nothing.'

He spat eloquently, and began to rub down his horse.

'The bitch,' I said, using Slaughterer's Spanish word which, when spoken with venom, richly expressed disappointment.

Comelli grinned, his teeth showing white in his half-grown black beard.

'You can search,' he said, 'but unless you have luck, it is no good. Last year when I was in the monte back there, I found many many holes of tatu carreta. I had nothing better to do, so I started to hunt for him. I just wanted to see what he looked like. Every night for a month I searched with my dogs but we could never get a single lousy scent. So I said to hell with it. One evening, three days later, when I had forgotten all about the rotten thing, a young one walked across the path in front of my horse. I just jumped off and caught him by the tail. It was not difficult, it was lucky. He was the first and the last one that I've seen.'

377

I had not been so recklessly optimistic as to imagine that Comelli would return with a giant armadillo, alive and kicking, tied across his saddle, but I had cherished the wisp of a hope that he would find a hole, some spoor, some droppings, anything to show that the creature inhabited some particular area of monte. With such evidence we could organize a careful and thorough hunt. Without it, a search was pointless.

Comelli smacked his horse on the rump and sent her away to graze.

'Do not be sad, *amigo*,' he said. 'You cannot tell with old man carreta. He might walk into the camp tonight.' He untied a cloth bag from his pack. 'Here. Perhaps this will make you a little happier.'

I unfastened the mouth of the bag and cautiously peered inside. At the bottom I saw a large ball of reddish fur.

'Will it bite?'

Comelli laughed and shook his head.

I reached inside and extracted first one, then two, and eventually four small furry kittens with bright eyes, elongated mobile snouts and long tails, ringed with black. They were baby coatimundis. For a moment my disappointment about the giant armadillo was lost in the pleasure of handling these delightful little creatures. They were quite fearless and clambered all over me, uttering little chirring growls, biting my ears and poking their noses into my pockets. They were so active that I could not hold them for long and soon they were on the ground scampering after one another, rolling over and over and chasing their own tails.

A full-grown coatimundi is a very formidable creature with huge canine teeth and a seemingly overwhelming desire to bite almost anything that moves, whether it be large or small. They wander through the bush in family groups, terrorizing the smaller inhabitants and devouring grubs, worms, roots, young nestlings, anything which is edible. Comelli's dogs had come across a female with a litter of ten. They had given chase and she had run up into a tree and, as the youngsters followed her, Comelli had managed to catch these four. They were still young enough to be tamed and there are few creatures more entertaining than a tame coatimundi. I was delighted with them.

We built them a large enclosure of saplings, woven together with creepers, into which we put a branch to serve as a climbing playground. For their first meal we gave them the only food we had available, some boiled mandioca. They fell upon it with enthusiasm, champing it noisily in their small jaws. Soon they stuffed themselves so full that they could no longer scamper but only waddle. They settled themselves in a corner, spent a few minutes scratching their distended stomachs, and then, one by one, they went to sleep.

But mandioca was not, by itself, a satisfactory food for them. They really needed meat. And so did we. We had had none for several days, but now, one way or another, we must get some. Before we reached a decision on exactly how we should do so, our problem was providentially solved for us.

Twenty peones galloped into Paso Roja, whooping and yelling and driving in front of them a bellowing steer.

'*Portiju,*' cried Comelli, scrambling to his feet and picking up a knife.

I had always imagined that if I were compelled to enter a slaughterhouse I should become a confirmed vegetarian over-night. But when the steer was lassoed within fifty yards of our camp, I was feeling so hungry that I watched it being slaughtered and butchered without any qualms whatever.

Comelli returned, his hands and arms bloodied to the elbow, carrying half the ribs of the steer over his shoulder. Within minutes they were sizzling and browning over the fire and, only three-quarters of an hour after the peones had first appeared, we were eating our first meat for many days. Knives seemed superfluous. We held the huge bow-shaped bones in our hands and picked off the tender meat with our teeth. I could not understand why it was so much more succulent and tender than the leathery beef that Elsita had produced for us.

'There's only two ways to eat Chaco beef,' said Sandy, in between mouthfuls. 'Either after it has been hung for several days, or like this – immediately after it has been killed and before rigor mortis has set in. And of the two, this is the best.'

I had to agree with him. Never had beef tasted so good.

The peones had come from an estancia many miles away and were scouring the Chaco for cattle that had strayed from their herds. Every few days they killed a steer for food and we were fortunate that they had decided to do so when they spent the night in Paso Roja.

The bonanza benefited everybody, for even such huge eaters as the cattlemen could not consume a whole cow by themselves. Slaughterer walked by, carrying a leg dripping blood. Patron and Bottle-opener had a flank between them. Comelli's dogs gobbled the offal and the coatimundis squabbled over pieces from the ribs. Black vultures had assembled in the trees overhead and were patiently awaiting an opportunity to descend on the remains of the carcass and claim their share.

Soon the ground between our camp and Patron's house was dotted with small crackling fires around which the peones sat in groups of twos and threes, and the whole monte was filled with the rich fragrant smell of roasting beef.

Comelli had managed to secure more than the ribs – he had also brought back a huge piece of shoulder muscle. We could not eat this straightaway so we decided to make it into *charqui* by cutting it into long strips and hanging it from a line to dry in the sun. Charqui, properly prepared, is not particularly appetizing but it will keep for quite long periods and still remain edible. No sooner had we finished preparing it and had returned to our fire, than a flock of Quaker parakeets flew on to the strips of meat and noisily began to gorge themselves. Parakeets, of course, are supposed to live on fruit and seeds, but these Quakers of the Chaco, living in such a barren country, had obviously learned to eat almost anything that was available. Nor were they the only birds to have modified their appetites in this

way, for soon they were joined by handsome red-headed car-
dinals, by saltators – large finches with black cheeks and orange
bills – and by mockingbirds, which flicked their long tails up
and down in order to maintain their balance on the string as
they pecked at the raw meat.

The Quaker parakeets eating drying meat

Comelli planned to continue the search for tatu carreta by a
reconnaissance westwards and I wanted very much to go with
him. All of us could not go, abandoning the oxen, our belongings and the coatimundis, but neither could Comelli and I take
both the horses and leave Charles and Sandy with none. We
then discovered that Bottle-opener had a spare horse. He said
that he did not want to lend it to us but he intimated that he
might be persuaded to sell it. Pancho, the horse in question, was
produced for our inspection. I am not skilled in the art of judging
the finer points of a horse, or of assessing its age by its teeth,

but it was quite clear, even to my inexperienced eye, that Pancho was exceedingly elderly. His cheeks were sunken, his back sagged in a deep arch, his ears flopped sadly and his head drooped. It occurred to me that Bottle-opener did not want to lend him to us in case the poor creature died from unaccustomed exertion, which he would doubtless be in danger of doing if he was ridden in the wild and ferocious manner affected by the peones. But I had no such intention. All I required was something that would amble gently onwards with me on its back and Pancho seemed capable of doing that. A young frisky creature would have been a positive embarrassment. Even so, I was not sure that I wished to buy Pancho.

'How much?' I asked.

'Five hundred guaranies,' replied Bottle-opener positively.

This was equivalent to about thirty shillings. Even Pancho, I thought, was worth that, and I bought him.

The next day, Comelli and I left together, taking with us a bag of farinha and some of the charqui. We followed a narrow track through the bush which here was higher and not as universally spiny as the monte around Estancia Elsita. The dogs ranged silently ahead of us, occasionally coming back to Comelli and then leaving again on another exploration into the undergrowth on either side.

In the late afternoon, Comelli suddenly reined his horse to a standstill and vaulted to the ground. By the side of the track yawned a gigantic hole. He did not need to tell me what it was. I knew at once, both from the triumphant look on his face and from the hole itself, that at last we had found the burrow of a giant armadillo. Fully two feet across, it had been dug in the side of a vast mound of compacted earth that was the nest of a colony of leaf-cutting ants. Great clods were scattered in front of it, some bearing deep grooves that had been made by the armadillo's massive front claws. I lay on my stomach and peered into the hole. Mosquitoes hummed in a cloud just inside the mouth. It went down farther than I could see, so I cut a stick from the bush and poked it inside. The tunnel was not a long one – no more than about five feet. It was not the permanent home of a tatu carreta, but merely a pit that it had dug into the side of the ants' nest in order to feed on them. Comelli meanwhile was following the trail that the animal had made when it left the hole. It led through the bush, round the ants' nest to the other side, where we found another similar hole that the creature had dug in its search for food. Its size and that of the lumps of earth that had been thrown out of it were dramatic testimony of the animal's great bulk and strength. Excitedly we followed the trail, cutting our way through the thorny undergrowth. Twenty yards away, we found a third hole. After half an hour of searching, we had discovered fifteen. The dogs confirmed that all of them were empty. None of them was anything more than an excavation for food.

We sat down to discuss the situation.

'The tracks could not have been made more than four days ago,' said Comelli, 'for the rains that fell on the day we arrived in Paso Roja would have washed them away. But they are not fresh. There is no scent and they are rather blurred. I think they are about four days old. Carreta is probably miles away by now.'

Even though this was disappointing, I could not but feel elated for at last I had seen tangible evidence of this beast which I had been beginning to suspect to be almost mythical. We searched painstakingly through the bush for some clue which would tell us in which direction the giant armadillo had departed. We found nothing and the trail was so old that the dogs were unable to find any scent to follow. All we could do was to continue westwards along the track and hope that somewhere we would find fresher traces of the animal.

At sundown we stopped. Comelli built a small fire and we made an evening meal of charqui.

'We will sleep a little until the moon rises,' Comelli said, 'and then we will start again.'

I spread my poncho by the fire, shut my eyes and dreamed

of the giant armadillo walking across the track in front of Pancho's feet.

When Comelli woke me, the moon had risen, white and full, and was illuminating the monte with a light so bright that it would have been possible to have read a book. Once more we saddled the horses and set off quietly through the bush. There was little to be heard but the jingle of harness and the sudden brushing of branches against our legs and the horses' flanks. From a distant part of the monte came the deep sombre hoots of an eagle owl, and in the ground beneath us crickets stridulated frantically until they were silenced by the thud of Pancho's hooves.

It was nearly midnight when suddenly we heard Diablo's high-pitched screaming yelps. He had found something. Even Pancho seemed to sense the excitement for, when I urged him towards the place where Diablo was baying, he broke into a canter and plunged boldly through the thorny bushes. I reached the dog at the same time as Comelli. Together we jumped down and forced our way into the thicket where he sat. He was crouching, snarling and yelping over an animal. Comelli called him off. There, lying on the ground, we saw an armadillo. It was a nine-banded one.

The dogs found two more nine-banded armadillos that night, but that was all. We made camp at three o'clock in the morning and slept until dawn.

We continued the search for three more days and nights. During the daytime it was extremely hot. We had long since drained the boiled muddy water that we had brought with us in bottles from Paso Roja and there were no waterholes or streams from which we could replenish them. When my thirst became insupportable, Comelli showed me how I could quench it even in this dry country. Low stumpy cacti were abundant in the bush and a section from one, trimmed of its spines, was refreshingly juicy. It tasted something like cucumber, but I was not very fond of it for it had an unpleasant aftertaste which set my teeth on edge. However there was another plant which provided a purer-tasting drink in even greater abundance. It was difficult to recognize, for it had only a small twiggy stem and sparse nondescript leaves, but two feet below ground its root was swollen to the size of a large turnip. The flesh of this distended root was white and translucent and so loaded with liquid that we obtained cupfuls merely by squeezing pieces of it in our fists.

Barren though the bush appeared to be, Comelli found much in it to supplement our meals of farinha and charqui. He cut the white centre shoots from the low carandilla palms which, he said, were particularly sought after by Amerindian women when they were suckling babies because they were so nourishing. They were white and nutty with a pleasant taste reminiscent of chicory. He showed me which berries were edible and which were poisonous. Once we found a fallen tree in which a colony of bees had made their nest. As Comelli prepared to chop it open, I suggested that we should build a smoky fire in order to drive most of the bees away and minimize the chances of our being stung. He was vastly amused at the idea. There were bees in the Chaco, he said, which did have painful stings but these were not they and though they buzzed alarmingly around our heads as we chopped at the trunk with our machetes, they made no attempt to molest us. We extracted the large honey-laden combs and ate them as they were – wax, pollen bread, grubs and all, the liquid honey dribbling down our chins.

Although we rode constantly throughout the day, we had little expectation of finding the giant armadillo itself for Comelli was sure that it seldom came out of its hole except at night, but we did hope that we might find some signs of the animal's presence. We discovered several more pits, but they were all of about the same age as the ones we had first found and none of them was a nesting hole. At night we relied on the noses of the dogs to detect the presence of any animal that was abroad. They found a hairy armadillo, several three-banded ones, and on one evening they caught and ate a fox. But neither they nor we found any recent traces of tatu carreta.

We travelled along the track until it emerged into the open plains. Comelli was positive that tatu carreta seldom if ever ventured beyond the shelter of the bush, so there was no point in going any farther. Sadly we returned to Paso Roja.

Charles and Sandy came out to meet us. As we sat round the fire telling them of what we had done and seen in the monte, Slaughterer walked into the camp. In his hand he held a large fluffy owl chick with huge yellow eyes, extravagantly long eyelashes and very large feet. Slaughterer grinned a little sheepishly as though he were really rather ashamed at being associated with such a childish thing. It was no part of his nature to be kind or considerate towards such insignificant beings as baby birds, but he did not dare to treat the little creature in too offhand a manner for it was quite obvious that he was hoping to sell it to us.

He placed the chick on the ground and joined us by the fire. The little bird stood upright, clopped its beak and trilled softly to itself. I pretended I had not noticed it.

'Good evening,' said Slaughterer punctiliously.

We responded with equal politeness.

He jerked his head towards the chick.

'Very good,' he said. 'Very rare.'

I laughed disbelievingly. 'That is a *ñacurutu* – an eagle owl. They are not rare.'

Slaughterer looked affronted.

'Very valuable bird. Much rarer than tatu carreta. I was going to keep him.'

He waited for me to express disappointment. I looked at the fire.

'I will let you have him if you would like him.'

'And what do you want to sell him for?'

This was the moment for which Slaughterer had been waiting, but now that it had arrived he seemed almost ashamed to put into words what I knew he had in his mind. He poked the fire with a stick.

'Your knife,' he mumbled.

We would soon be leaving the Chaco for good and I could easily buy another knife in Asunción. I handed it over to Slaughterer and took the young chick to give it a meal.

The baby owl was our last acquisition in Paso Roja. The next morning we had to leave, for in five days' time a plane was coming to Estancia Elsita to take us back to Asunción.

30

Moving a Menagerie

The weather, which had been hot and rainless for over two weeks, changed once more as we rode into Estancia Elsita and the clouds which had been gathering in the sky since the scarlet dawn burst into a heavy storm. Within a few hours the airstrip by the house was waterlogged. That night Faustino spoke to Asunción Airport on the radio and cancelled the flight of the plane which was due to collect us the next day.

It was nearly a week before he was able to call them again and report that the airstrip had dried out sufficiently to allow an aircraft to land in safety.

At last, however, the plane arrived. Carefully we stowed inside it the armadillos, the caiman, the coatimundis, the young owl and all the rest of the animals. As we gathered on the airstrip making our last farewells, Spika appeared with three baby parakeets. To his immense satisfaction he completed a final sale. Faustino gave us a huge side of raw beef which he asked us to deliver to his relations in Asunción. 'The unhappy ones,' he said. 'They are never able to get good Chaco beef.' Elsita, still chewing a cheroot, brought out the babies to watch us depart. When Comelli said goodbye, he shook me warmly by the hand. 'I will go on looking for old man carreta,' he said, 'and if I find him before you have left Paraguay, I'll ride to Asunción and bring him to you myself.'

The plane roared into life and we slammed the door. Two hours later we were in Asunción.

We were overjoyed to find that all the animals we had collected on the Curuguati and at Ita Caabo had flourished under Appolonio's devoted care. Many of them had grown beyond recognition and Appolonio had added to their number some opossums and toads which he had caught himself.

Now began the busiest and most worrying period of the whole expedition. All the animals had to be rehoused in light travelling cages. Customs officers had to inspect and count them. Officials from the Ministry of Agriculture had to examine them and certify that they were in good health and free from infectious diseases. Detailed arrangements had to be made with the airlines to fly the menagerie and ourselves on freighter planes down to Buenos Aires, thence to New York and finally to London, and the numerous regulations concerning the transit of animals at our various ports of call had to be disentangled from official documents and carefully studied to make sure that we had all the necessary papers and health certificates.

At the same time as doing all this, we had to feed and clean the animals themselves. Even with Appolonio doing much of the work, this seemed like a full-time job in itself. The babies were by far the most trouble. The owl chick could not be given ordinary meat, for all owls require bits of fur and gristle, sinews and feathers, which they regurgitate as pellets. If they do not get these ingredients in their food then their digestion seems to go wrong. Consequently Appolonio and his brother, the gardener, spent a great deal of time catching rats and lizards which we had to chop up and feed to the chick by hand. We also had some toucan nestlings which were incapable of feeding by themselves and three times a day we had to give them berries and small pieces of meat which they demanded should be pushed deep into their grotesque beaks and halfway down their throats.

When we had first arrived in Paraguay, I had confirmed that the only practical way of flying a cargo of animals from Asunción to London was through the United States. It was a long way round, there was the possibility that we might be delayed in making our connections and, as it was now December, we should be faced with the problem of finding heated accommodation for the collection during the time we should have to spend in New York. It was not an ideal route but we believed it to be the only one.

Then Sandy told us that by chance he had met the local representative of a European airline who had claimed that he could quite easily arrange for us to fly direct from Buenos Aires to Europe, thereby reducing the journey by many hours. This would be a far more satisfactory arrangement and we rushed round to the airline office to find out the details. Sandy's friend confirmed that this was indeed possible, for, he said, although there were no freighters crossing the Atlantic from Buenos Aires many of his company's passenger planes were returning from South America three-quarters empty at this time of the year and he was certain that he could get special permission for one of them to take us and our menagerie. All that he required was a list of the animals. With alacrity we produced a copy of the exhaustive catalogue we had prepared for the Customs authorities which listed the sex, size, and age in years and months of every animal we possessed.

He read it through out loud, in a wondering tone of voice. When he came to the armadillos, his brow furrowed and he reached down a bulky manual of regulations. After studying the index for some considerable time, he looked up at us.

'What are these animals, please?'

'Armadillos. They are rather charming little creatures, actually, with hard protective shells.'

'Oh, tortoises.'

'No. Armadillos.'

'Maybe they are a kind of lobster.'

'No, they are not lobsters,' I said patiently. 'They are armadillos.'

'What is their name in Spanish?'

'Armadillo.'

'In Guarani?'

'Tatu.'

'And in English?'

'Strangely enough,' I said jocularly, 'armadillo.'

'Gentlemen,' he said, 'you must be mistaken. There must be some other name for them, because armadillo is not mentioned in the regulations and *all* animals are listed in here.'

'I am sorry,' I replied, 'but that is their name and they have no other.'

He shut up his book with a bang.

'Never mind,' he said gaily, 'I will call them something else. I am sure it will be all right.'

On the strength of his assurances, we cancelled the elaborate arrangements we had made to travel via New York.

Two days before we were due to leave Asunción, the airlines man reappeared at the house with a worried look on his face.

'I am very sorry,' he said, 'but my company cannot accept your cargo. Head office in Buenos Aires say that those animals that were not mentioned in the manual will smell too bad.'

'Nonsense,' I said indignantly, 'our armadillos do not smell at all. What did you say they were?'

'I just called them something which I was sure that no one would have heard of before. I could not remember the name you said, so I found one in my son's animal book.'

'What did you say they were?' I repeated.

'Skunks,' he replied.

'Please,' I said, trying to contain my fury, 'will you cable Buenos Aires and explain that my armadillos are not skunks. They do not smell. Come and see for yourself.'

'It is no good now,' he said contritely, 'the space has been reserved for another cargo.'

That afternoon we had to return to our original airline and, full of apologies, asked if it would be possible to renew all the arrangements to travel by way of New York, that a week ago we had cancelled.

The longer we stayed in Asunción, the larger our problems became, for by now the whole of Paraguay seemed to have learned of our existence and men from all over the country began to converge on our house on bicycles, in rattling lorries and on foot, bearing a wide assortment of animals in gourds, boxes and stringbags. The rarest and most exciting of these last-minute acquisitions was brought to us by a man I had met in the Concepción hotel when Sandy and I had gone there on the first of our abortive sorties in pursuit of the giant armadillo. He arrived at the house trundling a handcart. On it, surrounded by a frail network of laths and string, stood a huge and extraordinary-looking wolf, a majestic creature, with a long reddish coat, large furry triangular ears, a white bib and fantastically elongated legs out of all proportion to the rest of his body. It was as though the image of a rather good-looking Alsatian dog reflected in a fairground distorting mirror had suddenly come to life. This was the rare *aguara guazu*, the maned wolf, which lives only in the Chaco and the northern part of Argentina. Its long legs enable it to run extremely swiftly and some people have claimed that it is the fastest of all land animals, excelling even the cheetah. Why it should require such speed is a mystery. There can be nothing from which it needs to escape – jaguars do not live on the open plains frequented by the wolf – neither is such extreme swiftness essential to catch the armadillos and small rodents on which presumably it preys, and there is no record of it ever attacking rheas which are the only things it might meet which could rival it in speed. It has been suggested that its height enables it to see for great distances over the flat plains and this is certainly true, but it hardly seems sufficient justification for the development of such extraordinary physique.

I was overjoyed to have it, for we had only just received a cable from the London Zoo saying that they had acquired from a German zoo a large male maned wolf, and asking if we could possibly find a mate for him. The one we now possessed was fortunately a female.

Housing her presented us with a great problem. Not only was her present cage so flimsy that it was quite insecure, but it was also so small that the poor creature was unable to turn round. Although her owner had told us that she was newly caught, she seemed quite docile and raised no objection when Appolonio and I fitted a leather collar around her neck. Cautiously we led her out of her cage and tethered her to a tree. I offered her some raw meat, but she spurned it. Appolonio insisted that we should give her some bananas. It seemed an unlikely diet for a wolf, but to my surprise she ate four immediately. After some time, she began tugging at her lead so persistently and energetically that I was afraid that she might injure her neck, so we shut up the kitchen chickens in their house and released her in the vacated hen-run. Then we set to work with saws and hammers to transform a large wooden crate into a cage for her. We finished it by the evening and put it in the chicken-run close by the wire. Coaxingly, we tried to persuade her to enter it, but she snapped and growled at us in a frightening manner. We changed our tactics. Appolonio put more bananas in the far end of the cage and sat himself in a strategic position on the other side of the wire, ready to drop the door behind her as soon as she ventured inside. Meanwhile I began work on a travelling cage for the coatimundis.

Dusk came and still the wolf showed no signs of entering her box. I walked over to consult with Appolonio and as I did so the wolf suddenly bolted and with a leap and a scramble she cleared the chicken netting and was gone.

The garden itself was securely fenced to keep out stray dogs so I was reasonably hopeful that it would prevent her escaping into the town, but the grounds of the house were immense and heavily planted with clumps of bamboo, flowering trees and decorative thickets of cactus. By now it was dark. We ran for torches and for an hour Charles, Appolonio and I searched the garden. We could find no trace at all of the wolf. She seemed to have disappeared entirely. We separated and each of us combed one section of the garden.

393

'*Señor, señor,*' shouted Appolonio from the other end of the garden. 'She's here.'

I ran across to him and found him shining his torch on the wolf which was sitting snarling in the middle of a small clearing surrounded by low cactus. Now that we had found her, I wondered rather vaguely what we did next. We had neither ropes nor nets nor cage. While I was still thinking, Appolonio leaped over the cactus and grabbed her by the neck. I could hardly hang back when he had so courageously shown the way, so I jumped over the cactus, dived at the struggling yapping pair and caught Appolonio neatly around the waist. By the time I had disentangled myself from him, the wolf had fastened her jaws on his hand so that I was able to straddle the animal and securely grip her head without any danger of being bitten myself. The wolf, feeling herself held from behind, released Appolonio's hand. To my relief, he had not been badly bitten. While all this had been going on, Charles, very sensibly, had gone to fetch the cage. After what seemed like an interminable delay, with the wolf struggling frantically in our arms, he arrived with it and we were able to bundle her inside.

At last all the arrangements were completed and the time came for us to leave Paraguay. Many of our friends came down to the airport to say goodbye and we took off for the last time from Asunción airport with feelings both of reluctance and relief.

We had two days to wait in Buenos Aires, but we managed to quarter the collection in the Customs shed and so avoided the complications of immigration and quarantine. While we were there, I heard that a friend of mine and his wife were also in the city at the beginning of their own animal-collecting expedition. I discovered his telephone number and rang him up. His wife answered the telephone and, after we had told her what animals we had managed to collect, she told me of their plans.

'Oh, by the way,' she said nonchalantly, 'we have got a giant armadillo.'

'How wonderful,' I said, doing my best not to sound jealous.

'Would it be possible for us to see it? We searched for one for so long in Paraguay. I would like to see what they actually look like.'

'Well,' she said, 'we haven't actually *got* it. But we have heard of a chap five hundred miles away in the north of Argentina who has caught one and we are going up there to collect it.'

I thought that it would be unfairly discouraging to relate the story of our experiences in Concepción. Months later I discovered that they were just as unlucky as we had been.

The departure of our freight plane was delayed by several hours and as a result we missed our connection in Puerto Rico. Fortunately there happened to be a luxurious passenger aircraft returning completely empty to New York and the airline authorities kindly allowed us and our animals to travel in it. We were now running short of animal food, but the steward on the plane had large supplies of unclaimed packaged dinners. I did not experiment to see if any of our animals would enjoy caviar, but the armadillos and coatimundis relished some smoked salmon, and the parrots dined eagerly on fresh Californian peaches.

As we landed in New York, I was alarmed to see that the ground was covered in snow. If we could not find a heated room for the animals within minutes of landing they would surely die. But I had forgotten the American passion for central heating. The animals were taken to an ordinary warehouse the inside of which seemed to me to be even hotter than an average day in Asunción.

The next night we were in London. Officials from the Zoo met us with heated vans and the whole collection was whisked off to Regent's Park. As they disappeared into the night, a load of worry dropped from my mind and was replaced by a feeling of relief that, throughout the six days that had passed since we had left Asunción, not one of the animals had shown any sign of illness or discomfort, and not one of them had died.

I went to see them at the Zoo many times in the weeks that followed. The eagle owl chick was almost fully fledged and had grown enormously. The Zoo already possessed a male eagle owl that had been without a companion for some years. Female owls are larger than their mates and, young though our bird was, she was already as big as her companion-to-be and well able to take care of herself when she was put in the same cage.

I was particularly anxious to see what happened when we introduced our maned wolf to the male that was already established and settled in the Zoo. One of the most important functions a zoo can perform is to establish breeding pairs of rare animals so that if the species is faced with extinction in the wild state it can be preserved in captivity and later, perhaps, zoo-bred animals may be released in reservations and re-established in their homelands. Ambitious though this may sound, the London Zoo has already played an important part in doing just such a thing. The rare Père David's deer, which once lived in China, but which became extinct there many years ago, was preserved in paddocks in the Zoo and at Woburn Abbey, the Duke of Bedford's home, and recently deer from the Zoo have been sent back to China to settle in the country where they have been extinct for half a century.

In time to come, the maned wolf may also be in danger of extinction. Already it is very rare, and year by year more of the Chaco is colonized by ranchers and brought under control. It was very important to us, therefore, that our female wolf and the male should accept one another. There was, nonetheless, a considerable risk in putting them together, for they might well fight and injure one another before they could be separated. Desmond Morris, then Curator of Mammals, and I watched as the keeper opened the gate which allowed the male wolf to walk into the same pen as the female. He trotted briskly out, but as soon as he saw her, he sprang back and stood taut and stiff, his mane bristling, his lips drawn back in a soundless snarl. She reacted in a similar way. Suddenly he snapped at her, but his jaws did not touch her. She snapped back at him and for a few seconds they sparred. They separated. The male slowly advanced upon her, his head held low. She stood her ground and submitted to his sniffs. Then she walked away and settled unconcernedly in a corner. He followed her. Soon the two were lying side by side, the male uttering gentle crooning noises deep in his throat and stroking her outstretched forelimbs with his front leg. There was no doubt; they had accepted one another. Perhaps, in years to come, there might be a family of these wonderful creatures in London.

Desmond Morris was very complimentary about our armadillos. We had brought fourteen of them of four different species, but I was still very sad that we had not managed to bring back a giant. I described to Desmond the huge holes that we had seen and the lengthy and abortive journeys that we had made in search of the creature. Desmond was fascinated by my descriptions and agreed with me that it would have been most exciting just to have seen this miraculous creature. But he charitably minimized our failure. 'After all,' he said, 'you have brought us more armadillos, both in actual numbers and in species, than we have ever possessed at one time before, and the three-banded belongs to a sub-species that has never been exhibited here alive.'

A week later he telephoned me.

'Wonderful news,' he said excitedly. 'By an extraordinary co-incidence I have just had a letter from a dealer in Brazil who says that he has got a giant armadillo.'

'How marvellous!' I replied. 'Are you absolutely certain that it really is a giant, or that he is not just trying to discover how much you would pay for one, like my friend in Concepción?'

'Oh yes. He's a very reputable dealer and he knows what he is talking about.'

'Well, I do hope you are going to get it,' I said.

'I certainly am!'

A week later he telephoned me again.

'That armadillo has just arrived from Brazil,' he said, 'but I'm afraid you are going to be rather disappointed. He is just a rather large hairy armadillo like your Podju. You can enrol me as Vice-President in your Failed to Find the Giant Armadillo Club.'

Three months later he telephoned me once more.

'I thought you might be interested to know,' he said in a flat voice, 'that we've got a giant armadillo.'

'Ha, ha!' I said, 'I've heard that story before.'

'No, he really is here in the Gardens. I've just been looking at him.'

'Good gracious. Where on earth did you get him from?'

'Birmingham!' said Desmond.

I went round to the Zoo immediately. The armadillo had been sent to the Birmingham dealer from Guyana; he was the first of his kind ever to have arrived in this country alive. Fascinated, I examined him closely and he peered back at me from his tiny black eyes. Over four feet long, he had gigantic front claws and, unlike any of the armadillos that we had caught, he seemed to prefer to walk on his hindlegs, with his front feet only just touching the ground. The plates of his armour were large and distinct, but pliable, so that he appeared to be wearing a coat of mail. He ambled up and down his den, trailing his stout plated tail like some antediluvian monster. He was one of the strangest and most fantastic beasts that I have ever seen in my life.

As I looked at him, I thought of the German in the forests beyond Concepción; of the huge holes and footprints we had found at Paso Roja; and of the nights that Comelli and I had spent searching through the thorny moonlit monte of the Chaco.

'Nice, isn't he?' said the keeper.

'Yes,' I said, 'he's nice.'

Sir David Attenborough is a broadcaster and naturalist whose television career is now in its seventh decade. After studying Natural Sciences at Cambridge and a brief stint in publishing, he joined the BBC. Since the launch of his famous *Zoo Quest* series in 1954 he has surveyed almost every aspect of life on earth and brought it to the viewing public. His latest programme, *Planet Earth II*, was the most-watched nature documentary of all time.